Tremor

From Pathogenesis to Treatment

Tremor: From Pathogenesis to Treatment
Giuliana Grimaldi and Mario Manto

ISBN: 978-0-031-00499-5   paperback

ISBN: 978-0-031-01627-1   ebook

DOI: 10.1007/978-0-031-01627-1

A Publication in the Springer series
SYNTHESIS LECTURES ON BIOMEDICAL ENGINEERING #20

Lecture #20

Series Editor: John D. Enderle, University of Connecticut

Series ISSN
ISSN 1930-0328   print
ISSN 1930-0336   electronic

Tremor
From Pathogenesis to Treatment

Giuliana Grimaldi
University of Palermo

Mario Manto
Université Libre de Bruxelles

SYNTHESIS LECTURES ON BIOMEDICAL ENGINEERING #20

ABSTRACT

Tremor represents one of the most common movement disorders worldwide. It affects both sexes and may occur at any age. In most cases, tremor is disabling and causes social difficulties, resulting in poorer quality of life. Tremor is now recognized as a public health issue given the ageing of the population.

Tremor is a complex phenomenon that has attracted the attention of scientists from various disciplines. Tremor results from dynamic interactions between multiple synaptically coupled neuronal systems and the biomechanical, physical, and electrical properties of the external effectors.

There have been major advances in our understanding of tremor pathogenesis these last three decades, thanks to new imaging techniques and genetic discoveries. Moreover, significant progress in computer technologies, developments of reliable and unobtrusive wearable sensors, improvements in miniaturization, and advances in signal processing have opened new perspectives for the accurate characterization and daily monitoring of tremor. New therapies are emerging.

In this book, we provide an overview of tremor from pathogenesis to therapeutic aspects. We review the definitions, the classification of the varieties of tremor, and the contribution of central versus peripheral mechanisms. Neuroanatomical, neurophysiological, neurochemical, and pharmacological topics related to tremor are pointed out. Our goals are to explain the fundamental basis of tremor generation, to show the recent technological developments, especially in instrumentation, which are reshaping research and clinical practice, and to provide up-to-date information related to emerging therapies. The integrative transdisciplinary approach has been used, combining engineering and physiological principles to diagnose, monitor, and treat tremor. Guidelines for evaluation of tremor are explained.

This book has been written for biomedical engineering students, engineers, researchers, medical students, biologists, neurologists, and biomedical professionals of any discipline looking for an updated and multidisciplinary overview of tremor. It can be used for biomedical courses.

KEYWORDS

tremor, oscillations, generator, signal processing, Fourier analysis, nonlinear, inertia, sensors, accelerometer, electromyography, haptics, fuzzy logic, brain computer interface, movement disorders, essential tremor, Parkinson's disease, cerebellum, muscle spindles, synapses, neurotransmitters, clinical scales, neuroimaging, blood studies, drugs, animal models.

Contents

CHAPTER 1

Introduction

1.1 DEFINITION OF TREMOR

Tremor is defined as a rapid back-and-forth movement of a body part (McAuley and Marsden, 2000). Tremor is one of the most common movement disorders encountered in clinical practice and is a readily apparent motor phenomenon in most instances. It occurs both in normal individuals (the so-called physiological tremor) and as a symptom of a disorder, most often of neurological origin.

Pathological tremor is usually a rhythmic and roughly sinusoidal oscillatory movement. However, tremor is a nonlinear and nonstationary phenomenon. It is distinct from other involuntary movement disorders such as chorea, athetosis, ballism, tics, and myoclonus (see Table 1.1) by its repetitive and stereotyped feature (Bhidayasiri, 2005). The different tremors are grouped according to their frequency, amplitude, topographical distribution, and task or position-dependence. The most commonly used clinical classification is to distinguish tremor into *rest tremor*, *postural tremor*, and *kinetic tremor* (see also Chapter 4 for the clinical characterization).

1.2 PHYSIOLOGICAL TREMOR

In physiological tremor, two distinct oscillations (mechanical reflex and central neurogenic) are superimposed upon a background of irregular fluctuations in muscle force and limb displacements (Elble, 1996, 2003). Frequency studies show, in the majority of cases, similar frequencies on both sides.

The *mechanical reflex component* is the largest of the two oscillations. Its frequency is governed by the inertial and elastic properties of the body (Elble, 1996). Damped oscillations are generated in response to pulsatile perturbations, such as those produced by irregularities in motor unit firings (a motor unit includes a motoneuron and the depending muscle fibers, see Chapter 2) and by blood ejection during cardiac systole.

The frequency (ω) of these passive mechanical oscillations depends directly upon the stiffness (K) of the joint and inversely upon the inertia (I) according to the equation:

$$\omega = \sqrt{(K/I)}.$$

TABLE 1.1: Differential diagnosis of involuntary movements		
INVOLUNTARY MOVEMENTS	**DEFINITION/FEATURES**	**DISEASES COMMONLY ASSOCIATED WITH THE MOVEMENT DISORDER**
Tremor		
Rest	See text	Parkinson's disease
Postural		Essential tremor
Kinetic		Cerebellar tremor
Dystonia	Prolonged muscle contractions leading to abnormal postures; may be repetitive; twisting movements	Drug-induced
		Genetic
		Idiopathic
Chorea	Irregular; often hidden in voluntary movement; generates a dance-like movement	Huntington's disease
Athetosis	Continuous slow hyperkinesia of distal segments of limbs; causes an octopus-like movement	Stroke
Ballism	Fast and ample movement of proximal segments of limbs; gives a "throw away"-like movement; more severe in upper limbs	Stroke
		Inflammatory diseases
Tics	Fast and short hyperkinetic movements usually with a facial or head topography	Gilles de la Tourette syndrome
Myoclonus	Sudden, short (20–150 ms) movement; may cause a pseudorepetitive muscular contraction	Essential myoclonus
		Myoclonic epilepsy
		Symptomatic myoclonus

Consequently, tremor frequency will increase from proximal to distal segments. Physiological tremor of the elbow has a frequency of 3–5 Hz, wrist tremor 7–10 Hz, and metacarpophalangeal joint tremor 12–30 Hz (Elble, 2003).

The *central neurogenic component* of physiological tremor is invariably associated with the modulation of motor unit activity. Rhythmic motor unit activity is not just a simple passive response to sensory feedback, but is driving the limb oscillation. Moreover, regardless of their mean frequency of discharges, participating motor units are entrained at about 8–12 Hz. The frequency of the central neurogenic tremor shows no response to modification of inertia or stiffness and is independent of the length of the stretch reflex. For these reasons, the central neurogenic tremor is believed to originate from an oscillating neuronal network within the central nervous system (Elble, 1996).

The normal tremor behavior when an inertial load is added to the limb is a decrease of frequency, according to the equation $\omega = \sqrt{(K/I)}$, when there is no significant contribution from the stretch reflex or central oscillations. This is confirmed by surface or needle electromyographic (EMG) studies showing no rhythmic motor unit entrainment despite the rhythmic oscillations of the limb (Figure 1.1a). In some subjects, a prominent motor unit entrainment with and without mass loading is identified (Figure 1.1b). In pathological cases (for instance, essential tremor, see Chapter 5), lower- and higher-frequency oscillations can be identified. These last ones are associated with motor unit entrainment and do not decrease with inertial loading. For these reasons, high-frequency oscillations are interpreted as central neurogenic oscillations, whereas low-frequency oscillations correspond to the mechanical reflex resonance frequency (Elble, 2003).

In addition to the 8- to 12-Hz frequency band, frequency oscillations in the 15- to 30-Hz range and around 40 Hz are commonly found in the kinematic and EMG recordings in the upper limbs. Based on the fact that changes in the amplitude of specific frequency bands of tremor as a function of added mechanical load are informative of the central or peripheral modulation of tremor, some authors have studied the effects of increments of load on the intensities of the 8- to 12-, 20- to 25-, and 40-Hz neural rhythms (Vaillancourt and Newell, 2000). Their findings support the view that the 8- to 12-Hz oscillation resides within the central nervous system with its amplitude remaining independent of mechanical resonance oscillations and that the 20- to 25-Hz tremor oscillations are related to the mechanical properties of the finger. This means that the 20- to 25-Hz band is related to cortical activity, but its amplitude is modulated by mechanical reflex oscillations. Concerning the 40-Hz peak found in EMG recordings, it is likely originating in the central nervous system before being low-pass filtered by the muscle–tendon complex. The tendency of human limb segments to exhibit rhythmic oscillations around 40 Hz is observed in EMG recordings during strong voluntary contractions ("Piper rhythm"; Piper, 1907). Data of surface EMG traces simultaneously recorded with magnetoencephalographic signals (see Chapter 6) suggest that Piper rhythm of human muscles is linearly correlated with focal activity in the controlateral motor cortex both in

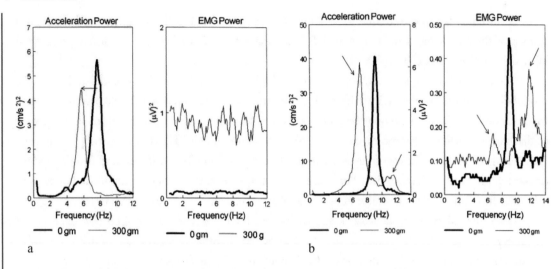

FIGURE 1.1: (a) Wrist tremor and rectified–filtered electromyography (EMG; Extensor carpi radialis bilateral: ECRb) from a 27-year-old woman with no evidence of motor unit entrainment without (thick trace) and with 300-g loading (thin trace). Note the 2-Hz reduction in the acceleration spectral peak with 300-g loading (arrow). (b) Wrist tremor and rectified–filtered EMG (ECRb) from a 21-year-old man with prominent motor unit entrainment with and without mass loading. With no mass loading, the acceleration and EMG spectra contained a single coherent peak at 9.4 Hz. Mass loading produced coherent peaks in the EMG and acceleration spectra at 6–7 and 11–12 Hz (arrows). Left vertical axis of the acceleration spectrum: 0-g load; right vertical axis: 300-g load. From Elble (2003), with permission from Elsevier.

isometric contractions and during phasic movements. A longer lag to tibialis anterior than that to forearm extensor muscle may be detected. This interval is linked to the conduction in fast pyramidal pathways (see Chapter 2). Because Piper rhythm can be picked up from most muscles, including those with few or no muscle spindles, it is unlikely that the coherence between cortical activity and the Piper rhythm is due to a simple reafference mechanism. Piper Rhythm is related to the degree of force exerted in tonic and phasic contractions probably because of a stronger excitation of the motor cortex during forceful contractions. Alternatively, it may be related to a greater attention demand in tasks requiring forceful contraction (Brown et al., 1998).

1.3   SOURCES OF TREMOR: NEURONAL NETWORKS

We can thus summarize the sources of tremor into three groups (Figure 1.2; Hallet, 1998):

I—Mechanical oscillations: Motion of joints and muscles obey the laws of physics, and tendon–muscle–joint complexes can be compared with masses and springs. Therefore, oscillations can be interpreted as related to movements of these masses and springs (Figure 1.3).

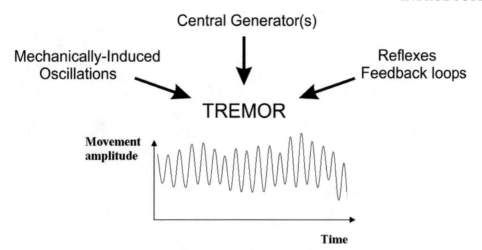

FIGURE 1.2: The three main sources of tremor.

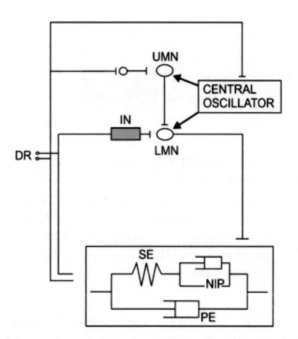

FIGURE 1.3: Central and peripheral loops in the nervous system. The figure illustrates the interaction between the central oscillator and the upper motoneuron (UMN)/lower motoneuron (LMN). IN indicates the pool of interneurons in the spinal cord. DR corresponds to the dorsal root ganglia. The rectangle in the bottom represents Hill's muscle model (SE: series elastic component; NIP: neural input processor in parallel with a viscous component PE).

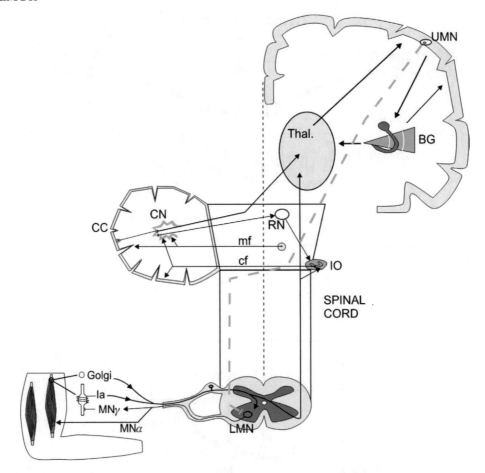

FIGURE 1.4: Pathways involved in tremorgenesis. Muscle spindles are receptors located inside muscles, made up by fibers sensitive to changes in length. In response to lengthening, the signals reach spinal and supraspinal centers where motor commands are generated and sent back to the extrafusal muscle fibers. UMN: upper motoneuron; Thal: thalamus; BG: basal ganglia; CC: cerebellar cortex; cf: climbing fibers; mf: mossy fibers; CN: cerebellar nuclei; IO: inferior olive; LMN: lower motor neuron; MNα: alpha motoneuron; MNγ: gamma motoneuron; RN: red nucleus; Ia: Ia sensory afferent fibers. Note that afferents project to the cerebellum ispilaterally.

II—Reflex oscillations: They are related to loops: *peripheral loops* from muscles to the spinal cord and back again (spinal level) and *central loops* from the periphery to the spinal cord and segments at the supraspinal level including the brainstem, cerebellum, basal ganglia, and cerebral cortex (Figures 1.3 and 1.4).

- The simplest loop is from the muscle spindle. Ia afferents are connected monosynaptically to the motoneuron, and the motor axon projects to the extrafusal muscle fibers.
- A classical example of a central loop is given by the role of comparator and movement controller of the cerebellum.

III–Central oscillations: Since the first electroencephalographic (EEG) recordings, it is obvious that the neural activity follows rhythmic behaviors. Cerebral cortex, basal ganglia, cerebellum, and brainstem nuclei are all involved in tremogenesis (see Chapter 3).

REFERENCES

Bhidayasiri R. Differential diagnosis of common tremor syndromes. *Postgrad Med J.* 2005;81: 756–762. doi:10.1136/pgmj.2005.032979

Brown P, Salenius S, Rothwell JC, Hari R. Cortical correlate of the Piper rhythm in humans. *J Neurophysiol.* 1998;80:2911–2917.

Elble RJ. Central mechanism of tremor. *J Clin Neurophysiol.* 1996;13:133–144. doi:10.1097/00004691-199603000-00004

Elble RJ. Characteristics of physiologic tremor in young and elderly adults. *Clin Neurophysiol.* 2003;114:624–635. doi:10.1016/S1388-2457(03)00006-3

Hallet M. Overview of human tremor physiology. *Mov Disord.* 1998; 3(3):43–48.

McAuley JH, Marsden CD. Physiological and pathological tremors and rhythmic central motor control. *Brain.* 2000;123(Pt 8):1545–1567. doi:10.1093/brain/123.8.1545

Piper H. Über den willkürlichen muskeltetanus. *Pflügers Gesampe Physiol Menschen Tiere.* 1907;119: 3001–3338. doi:10.1007/BF01678075

Vaillancourt DE, Newell KM. Amplitude changes in the 8–12, 20–25 and 40 Hz oscillations in finger tremor. *Clin Neurophysiol.* 2000;111:1792–1801. doi:10.1016/S1388-2457(00)00378-3

· · · · ·

CHAPTER 2

Anatomical Overview of the Central and Peripheral Nervous System

This chapter provides an overview of the relevant anatomy for the understanding of the organization of the central and peripheral nervous system. This anatomical framework is the substratum of tremor in human and must be kept in mind when assessing tremor, as several key features of tremor are directly explained by the organization of these anatomical structures. Moreover, several therapies [such as deep brain stimulation (DBS), see Chapter 8] are based on this organization.

2.1   CEREBRAL CORTEX

The cerebral cortex is the layer of gray matter in the cerebral hemispheres' external coat. The surface of the cerebral cortex is folded in sulci. The cerebral cortex is divided into four lobes: frontal, parietal, temporal, and occipital lobes. Two additional structures are included in the cerebral cortex: the insula and the limbic lobe.

Cortical areas are linked to subcortical structures through afferent and efferent fibers. Moreover, commissural tracts connect both cerebral hemispheres. From a functional point of view, cortical areas are divided into motor, sensory, and associative areas. According to Brodmann classification, each area is identified by a number (Figure 2.1).

1. Frontal lobe
    *   Area 4 corresponds to the frontal ascendant circonvolution. This primary somatomotor area issues motor commands along the pyramidal tract. Commands reach the brainstem nuclei and spinal cord. As a consequence of the somatotopical distribution of the neurons controlling the limbs, face, and trunk movements, a representation of the contralateral hemibody is discernible (the so-called motor homunculus). Motor areas receive main projections from the thalamus.
    *   Area 6, or premotor area, sends information mainly to Area 4, pontine nuclei, red nucleus, striatum, and mesencephalic substantia nigra.
    *   Area 8 allows conjugated eye movements in response to head rotation.

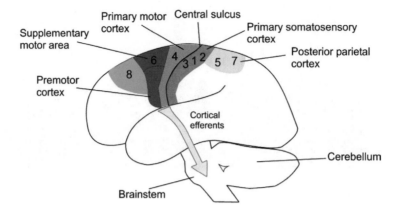

FIGURE 2.1: Representation of the brain. Main Brodmann's areas involved in tremor are illustrated.

2. Parietal lobe
   - Areas 3, 1, 2 are the somatosensitive areas. They are located in parietal ascending circonvolution, behind Rolando's scissura. A sensory homunculus is represented. These areas are highly connected with the motor areas, so that motor activity is always performed in a context of a sensory afferent according to the cortical reflex arch, except for deafferentation states.
3. Occipital lobe
   - Area 17 is the visual primary area. Vision plays an important role in the modulation of tremor intensity.

Associative areas integrate information from different systems. The temporo-parieto-occipital cortex is crucial for a complex sensorial elaboration that integrates somatic, acoustic, and visual inputs. The *prefrontal area* is involved in the planning of voluntary movements. The *posterior parietal area* coordinates somatic and visual inputs and integrates them into movement engrams.

2.2 BASAL GANGLIA AND THALAMUS

The basal ganglia are made up of five deep subcortical nuclei: caudate and putamen (forming the striatum), globus pallidus, subthalamic nucleus, substantia nigra, which is divided into pars reticulata and pars compacta (Figure 2.2). In each cerebral hemisphere, the basal ganglia receive information from the cerebral cortex and project it to the thalamus, prefrontal, premotor, and motor cortex. Therefore, motor functions of the basal ganglia are mediated partly by the frontal cortex. In addition to the motor functions, the basal ganglia also play key roles in cognitive operations, emotions, and learning.

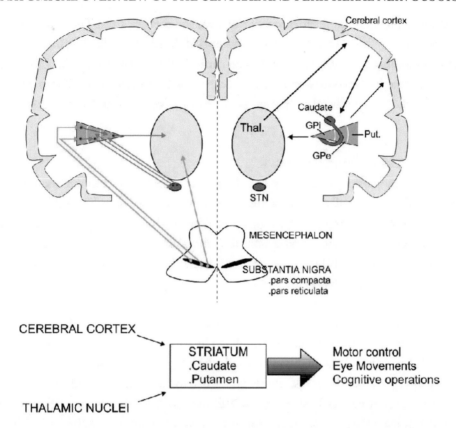

FIGURE 2.2: Coronal representation of the motor circuits of the basal ganglia. Main anatomical connections are illustrated. Nearly every input to the basal ganglia reaches the caudate nucleus and the putamen (Put.). These two nuclei form the striatum. The subthalamic nucleus (STN) receives projections from the external portion of the pallidum (GPe) and projects back to the globus pallidus and to the pars reticulata (substantia nigra). In addition, the STN receives projections directly from the primary motor cortex/premotor cortex, allowing a direct control to the output of basal ganglia. The striatum receives an important dopaminergic projection from the pars compacta. Thalamic nuclei are the main targets of the basal ganglia.

The striatum receives the majority of the afferents from the cerebral cortex (motor, sensitive, associative, and limbic areas) and the thalamus. Inputs from the cortex to the striatum are glutamatergic and therefore excitatory (see also Chapter 3). Projections maintain a topographical organization, so that putamen is principally involved in motor control, caudate in the ocular movement control and in some cognitive operations, and their ventral part in limbic functions. From the globus pallidus and substantia nigra pars reticulata, the basal ganglia fibers reach the thalamic nuclei [nuclei ventral lateral (VL), ventral anterior (VA), mediodorsal (MD), and centromedian (CM)], which project back to the prefrontal, premotor, supplementary motor area, and motor cortex. The globus pallidus and substantia nigra pars reticulata are the two main outputs of the basal ganglia. The basal ganglia influence motor descending systems (corticospinal and corticobulbar). The connection with the superior colliculus is involved in the control on ocular movements. All these interconnections give the basal ganglia an important role in the planning and performance of complex motor strategies.

2.2.1   Direct and Indirect Pathways

Projections from the striatum reach the internal segment of the globus pallidus and the substantia nigra pars reticulata and, finally, the thalamus. This is the *direct pathway*. The information leaving the striatum to the external segment of the globus pallidus and the subthalamic nucleus proceeds via the *indirect pathway*. Projections get back from the subthalamic nucleus to the internal segment of the globus pallidus and the substantia nigra pars reticulata. The neurotransmitters of the direct and indirect pathways are distributed as follows:

1. Direct pathway: neurotransmitters are GABA and substance P, which are inhibitory. Therefore, when the striatum is stimulated by cortical projections, it inhibits the globus pallidus and the substantia nigra pars reticulata. Given that these two nuclei have inhibitory effects on the thalamus, the thalamus is released from tonic inhibition, and movements are allowed. Thalamus–cortical activation induces, in turn, activation of premotor and supplementary motor areas projecting upon the motor cortex, brainstem, and spinal cord.

2. Indirect pathway induces a decrease of activation of the neurons of motor areas. Indeed, the striatum inhibits (via GABA and enkephaline) the external segment of the globus pallidus, which, in turn, inhibits the subthalamic nucleus. The subthalamic nucleus' excitatory influence on the internal segment of the globus pallidus and the substantia nigra pars reticulata allows the inhibitory action of these latter nuclei upon the thalamus.

Additional effects are provided from dopaminergic projections of the substantia nigra pars compacta, which has a facilitatory action on movement by exciting the direct pathway and inhibiting the indirect pathway.

The thalamus is thus a crucial synaptic relay both for sensory inputs and motor output. The thalamic nuclei retransmit motor codes—sent by cerebellar circuitry and basal ganglia—toward motor areas of the frontal lobe. Moreover, the thalamus is involved in the maintenance of consciousness and contributes to autonomic responses.

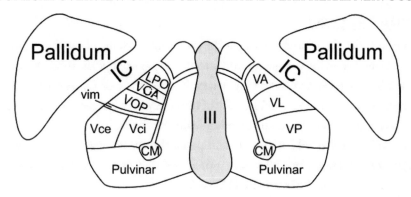

FIGURE 2.3: Illustration of the thalamic nuclei. Most commonly used terminologies are shown. LPO: lateral polaris, VOA: ventralis oralis anterior, VOP: ventralis oralis posterior, Vim: ventralis intermedius, Vci: ventralis caudalis interna, Vce: ventralis caudalis externa, CM: centrum medianum, VA: ventralis anterior, VL: ventralis lateralis, VP: ventralis posterior, IC: internal capsule, III: third ventricle. Adapted from Goldman and Kelly (1995).

2.2.2   Thalamic Nuclei

The thalamic nuclei can be divided into the relay nuclei and the diffuse-projection nuclei. Six groups of thalamic nuclei are described (Figure 2.3):

1. Lateral nuclei (ventral and dorsal tiers): receive restricted sensory and motor afferents and project to specific cortical regions (sensory, motor, or association cortex). They are made up of the following nuclei: *ventral anterior* (VA) and *ventral lateral* (VLN) involved in motor control; *ventral posterior* (VP) critical for somatic sensation; *medial and lateral geniculate* mediating information about hearing and vision, respectively; *lateral dorsal*; *lateral posterior*; *pulvinar* (the biggest thalamic nucleus). The pulvinar is involved in the integration of sensory information.

2. Medial nuclei: most representative is the *medial dorsal*.

3. Anterior nuclei: involved in emotions.

4. Intralaminar nuclei: diffuse-projection nuclei. The largest is the *centromedian nucleus* (CM) that projects to the frontal cortex and the basal ganglia (striatum).

5. Midline nuclei: diffuse-projection nuclei located in the dorsal half of the wall of the third ventricle.

6. Reticular nucleus: located in the lateral side of the thalamus. This nucleus is strictly interconnected with a specific relay nucleus and is the sole nucleus having inhibitory output and no projections towards cerebral cortex.

2.3 CEREBELLUM

The cerebellum overlies the posterior parts of the pons and medulla, occupying a large part of the posterior fossa. Structurally, the cerebellum consists of four pairs of nuclei embedded in white matter and surrounded by a cortical mantle of gray matter (Colin et al., 2002). Although the circuitry is functionally heterogeneous, the cytoarchitecture of the cerebellum is remarkably homogeneous. The cerebellum contains more neurons than any other region of the brain.

The cerebellum is divided into three main parts: the *anterior lobe*, the *posterior lobe*, and the *paraflocculus/flocculus* separated by the fissure prima and the fissure posterolateralis, respectively (Figure 2.4a). Three areas can be considered (Dow, 1942):

1. Flocculonodulus *(vestibulocerebellum)*
2. Vermal anterior and posterior lobes with mainly spinal connections (*paleocerebellum*)
3. Mediolateral part having principally corticopontocerebellar connections (*neocerebellum*)

Afferents enter the cerebellum through three pairs of peduncles: the inferior peduncle (or restiform body), the large middle peduncle (or brachium pontis), and the superior peduncle (or brachium conjunctivum). Efferents from the cerebellar nuclei leave the cerebellum through the superior and inferior peduncles. The number of afferent fibers exceeds the number of efferents by a ratio of about 40:1, hence the enormous computational capabilities of the cerebellar circuitry.

Three types of fibers enter the cerebellum:

1. Climbing fibers: thin, myelinated and slow conductive (20 m/s) fibers firing at low frequency (about 1 Hz). They arise exclusively from the contralateral inferior olive, crossing the midline, and ascending in the restiform body.
2. Mossy fibers: large myelinated fibers. They are fast conducting and fire at high frequency rates. They arise from a large spectrum of ipsilateral and contralateral sources.
3. Cholinergic and monoaminergic afferents: distributed diffusely.

2.3.1 Cerebellar Cortex

The cerebellar cortex (Figure 2.4b) contains *Purkinje cells*, *granule cells*, and *inhibitory interneurons* organized in a trilayer structure with the Purkinje cell layer (ganglionic layer) separating the outer molecular from the inner granular layer.

1. *Purkinje cells* are GABAergic and thus inhibitory (Ito and Yoshida, 1964). Their axons project to the cerebellar nuclei and vestibular nuclei. Purkinje cells receive a glutamatergic

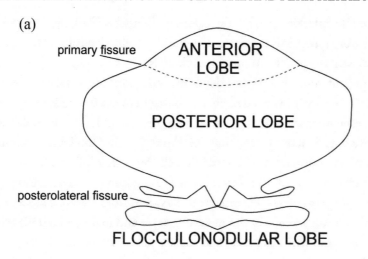

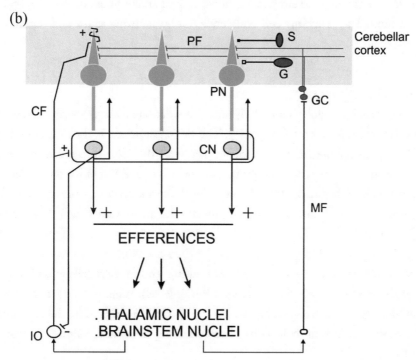

FIGURE 2.4: (a) Illustration of the unfolded cerebellum subdivided in three lobes. (b) Organization of the cerebellar cortex. Purkinje neurons (PN) project upon cerebellar nuclei (CN). Inhibitory interneurons of the cerebellar cortex (S: stellate cells, G: Golgi interneurons) modulate the activity of Purkinje cells. Granule cells (GC) are the source of parallel fibers (PF) branching upon Purkinje cells. Climbing fibers (CF) emerge from the inferior olive (IO). Mossy fibers (MF) emerge from various sources in the brainstem.

projection from the inferior olive. This synapse-climbing fiber/Purkinje neuron is the most powerful excitatory synapse of the brain. The Purkinje cell's dendrites also establish excitatory synapses with the parallel fibers of granule cells (Eccles et al., 1967).

2. *Granule cells* present about four to five dendrites and a thin unmyelinated T-shaped axon whose branches constitute the parallel fibers running between the Purkinje neurons.

3. Inhibitory interneurons are located both in the molecular and in the granular layer. They balance the excitatory activity targeting the Purkinje cells. Inhibitory interneurons in the molecular layer outnumber the Purkinje cells by a ratio of 10:1 (Andersen et al., 1992; Pouzat and Hestrin, 1997). Interneurons residing in this layer are the *basket cells* and *stellate cells*. *Lugaro cells* receive massive innervation from Purkinje cell axon collaterals. In turn, their axon ascends in the molecular layer and makes numerous symmetrical GABA-ergic synapses with the basket and stellate cells (Lainé and Axelrad, 1998). *Golgi cells* receive the main excitatory input from the parallel fibers on the dendritic arbor and are inhibited by Purkinje cell collaterals. Golgi neurons sense and regulate the mossy fibers' input.

2.3.2 Cerebellar Nuclei

Cerebellar nuclei (fastigial, globosus, emboliformis, and dentate nucleus) receive the inhibitory projection from the cerebellar cortex (Ito, 1984). The dentate nucleus is the largest and appears as a convoluted band, having the shape of a folded bag, with the hilus directed medially (Carpenter, 1985). Nucleofugal fibers are excitatory (glutamatergic) anywhere except for the nucleo-olivary projections (Schwarz and Schmitz, 1997). All nucleofugal fibers send collaterals that end in the granular layer. The nucleocortical projections reach mostly the areas from which they receive Purkinje cell axons (Buisseret-Delmas, 1988; Buisseret-Delmas and Angaut, 1988; Trott et al., 1998). This is one example of an internal loop within cerebellar circuitry. Nuclear cells receive a dense Purkinje cell innervation. On overage, the ratio between Purkinje and nuclear cells is 26:1, a given nuclear cell receiving about 14 terminals from the same Purkinje cell. Inhibitory inputs to the nuclear cell represent ~60% of the total synaptic input (Ito, 1984). On the other side, the excitatory input on nuclear cell comes, partly from collaterals of the climbing fibers and, extensively from the mossy fiber afferents.

2.3.3 Climbing Fibers

Arise from the olivary complex. This complex is organized in functional units that tend to fire synchronously. The anatomical and functional unit of the olivocerebellar system consists of a sagittal band of cortex receiving climbing fibers from a small part of the inferior olive and sending its output to a small part of the nuclei. A given area in the inferior olive corresponds to a definite area in the inferior output structure. The inferior olive translates its afferent inputs in to a well-controlled, low-frequency (1 Hz) discharge in time and distributed to a precise cluster of sagittally organized Purkinje cells in space. Inferior olive receives projections from the spinal cord, motor cortex, sensory spinal root of the trigeminal nerve, and red nucleus (Colin et al., 2002).

2.3.4   Mossy Fibers

Mossy fibers have numerous origins, mainly somesthetic, vestibular, acoustic, visual, and cortical. The principal spinocerebellar tracts are the dorsal and the ventral spinocerebellar tracts. The *dorsal spinocerebellar tract* (DSCT) arises from Clarke's column. It conveys information mainly from the hind limb and enters the cerebellum via the inferior peduncle. Its cells are either proprioceptive—stimulated by muscle spindles primary Ia and secondary group endings II, or by tendon Ib endings—or exteroceptive cells—monosynaptically activated by cutaneous low-threshold, fast-adapting hairy, and high-threshold, slowly adapting pressure receptors. The *ventral spinocerebellar tract* (VSCT) arises from the third to the sixth lumbar segments and conveys information from the hind limb. Axons cross the midline, ascend ventrally to the DSCT in spinal cord, cross a second time at the level of the ipsilateral brachium conjunctivum, and end bilaterally in the anterior lobe. All ventral spinocerebellar neurons receive strong polysynaptic input from the ipsilateral flexor reflex afferents (Oscarsson, 1973).

Fibers from vestibular nuclei project diffusely to the vermis, the flocculus, the paraflocculus, the paramedian lobule, the fastigial, and the interpositus nucleus, but not to the lateral cerebellum.

Afferents entering the cerebellum through the brachium conjunctivum emerge also from the reticular precerebellar nucleus. Fibers from lateral reticular precerebellar nuclei (located in the lower medulla) ascend bilaterally through the brachium conjunctivum (Clendenin et al., 1974, 1975; Ekerot, 1990a–c). Other groups of fibers—from the nucleus reticularis tegmenti pontis (NRTP)—relay visual information to the flocculus and are involved in the control of gaze.

Each large Brodmann's area projects to one or two slabs oriented rostrocaudally in the pons (Serapide et al., 1994). Each slab projects mainly contralaterally to the cerebellar hemisphere in a well-defined region.

2.3.5   Mediolateral Subdivision of the Cerebellum

Three zones have been identified: a vermal zone that projects to the fastigial nucleus, an intermediate zone whose Purkinje cells are connected with interposed nuclei, and a lateral zone projecting to the dentate nuclei.

2.3.6   Projections of Cerebellar Nuclei

The cerebellar ouput arises exclusively from the cerebellar nuclei, except for the vermal Purkinje cell axon to lateral vestibular nucleus.

The fastigial nucleus projects bilaterally to the vestibular and reticular nuclei. The fastigial afferents are involved in the control of reticulospinal and vestibulospinal tracts. Their effects on the spinal alpha and gamma motor neuron are complex (Ito, 1984).

The anterior interpositus nucleus projects mainly to the ventral lateral (VLN) and ventral anterior (VA) nuclei of the contralateral thalamus and sends collaterals to the caudal magnocellular part of the red nucleus (mesencephalon), pontine nuclei, superior colliculus. The posterior interpositus nucleus sends projections towards the thalamus (lower density of projections if compared with those from the anterior part) and to the medial part of the red nucleus. The dentate nucleus has similar projections of the interpositus. However, there are two differences: a small number of fibers from the dentate nucleus project to the rostral intralaminar thalamic nuclei, and the collaterals reach the anterior parvocellular part of contralateral red nucleus.

The thalamic nuclei are thus the major sites that receive cerebellar nucleofugal projections. In turn, thalamic neurons project to the cerebral cortex. Through this *thalamocortical relay*, the fastigial nucleus projects bilaterally to the hind limb area of the motor cortex and the parietal cortex. The interpositus nucleus projects contralaterally to the trunk area of the motor cortex and the premotor cortex. The dentate nucleus projects contralaterally to the forelimb area of the motor cortex, premotor cortex, and prefrontal association cortex.

2.3.7   Guillain–Mollaret Triangle (Dentatorubro-Olivary Pathway)

This anatomical notion refers to the projection from the dentate nucleus to the contralateral red nucleus and inferior olive, via the superior cerebellar peduncle, the central tegmental tract, and the inferior cerebellar peduncle. This pathway plays critical roles in tremor genesis.

2.4   SPINAL CORD

The spinal cord is a cylindrical-shaped portion of nervous matter located inside the vertebral bone canal. The rostral extremity is in continuity with the medulla. The caudal extremity forms the conus medullaris ending in the filum terminalis in correspondence to coccygeal bones. The spinal cord is organized into 31 continuous spinal segments divided into four regions: cervical, thoracic, lumbar, and sacral. Spinal nerves enter and leave the lateral medullary zones through the dorsal (sensory) and the ventral (motor) roots. Once compacted, the spinal nervous trunk gets off the vertebral canal through the intervertebral foramina. The area of skin supplied by axons from a single dorsal-root ganglion is defined as a *dermatome*.

In a cross-section, the spinal cord shows the butterfly-like shaped gray matter surrounded by the white matter.

1.   The medullar gray consists of cell bodies and dentrites of neurons and glial cells. The gray matter is subdivided into *dorsal or posterior horn*, made up of the sensory nuclei where somato-sensory information ascends toward the brain stem and thalamus, and in *ventral or motor horn* containing motor neurons innervating skeletal muscles. Information flows from dorsal horn

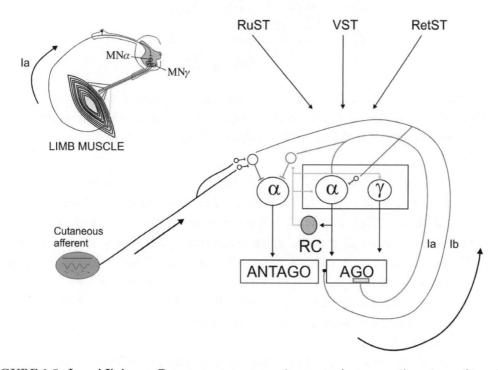

FIGURE 2.5: Ia and Ib loops. Gamma-motor neurons innervate the contractile regions of intrafusal fibers and regulate the sensitivity of muscle spindles; they are coactivated during motion. Ia afferent input (originating from a muscle spindle illustrated by a small red rectangle; spindles are arranged in parallel with muscle fibers) inhibits antagonist muscles via reciprocal inhibitory interneurons. Ib Golgi tendon organ afferents (the main source of autogenic inhibition) are arranged in series with muscle fibers. They are sensitive to changes in tension. Flexor motoneurons may receive excitation not only from antagonist extensor muscles, but also from extensors operating at other joints. Afferent fibers from Golgi tendon organs provide a negative feedback to the homonymous and synergic alpha-motoneurons and excite the antagonist motoneurons via an excitatory interneuron. Ib inhibitory interneurons receive conver-gent input originating from the skin and joints. α: alpha-motoneuron, γ: gamma-motoneuron; RuST: rubrospinal tract; VST: vestibulospinal tract; RetST: reticulospinal tract. VST stimulates directly alpha-motoneurons. Adapted from Manto et al. (2008).

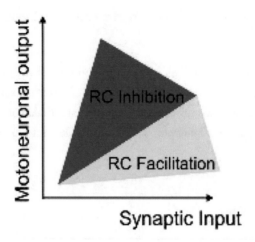

FIGURE 2.6: Renshaw cells (RC) modulate the motoneuronal output from the spinal cord (spinal recurrent inhibition). In case of RC facilitation via motor axon collaterals, the output from motoneuronal output is decreased compared to the state of RC inhibition. Adapted from Manto et al. (2008).

to motoneurons (or between different groups of motoneurons) and is modulated by numerous interneurons. In particular, the inhibitory Renshaw cells allow a regulation of motor neuronal activity (Figures 2.5 and 2.6). Between the two horns, the *intermediate zone* contains neurons whose axons project either in the ventral horn or in the brain stem and cerebellum.

2. The white matter consists mainly of axons grouped into tracts. Three large columns are found:
   - *Dorsal column*: composed by primary sensory fibers afferent to the brain stem
   - *Lateral column*: contains both axons ascending to the higher levels of the central nervous system and axons coming from brain stem nuclei and cortex that project on spinal motoneurons and interneurons
   - *Ventral column*: made up of fibers transmitting to the higher levels of the central nervous system pain and thermal sensations, as well as descending motor axons that control axial muscles and posture. Moreover, ventrally to the central canal, the *ventral commissure* is made up of packed axons that cross from one side to the other side of the spinal cord. Some of them transmit information about sensation of pain; others are involved in the control of posture.

The principal *spinal tracts* are the following:
1. Anterior column
   - *Longitudinal medial tract*: from mesencephalon to spinal cord. It represents the connection between vestibular nuclei and the nuclei controlling the eyes and head muscles

movements, allowing the maintenance of balance by the synergical movements of the eyes and head.

- *Anterior tectospinal tract*: part of the extrapyramidal pathways. Fibers start in the mesencephalic tectum (superior quadrigeminal tuberculi), cross the midline, and synapse with motoneurons at the level of the cervical anterior horn.
- *Vestibular spinal anterior tract*: made up of crossed and uncrossed fibers from the vestibular lateral nucleus of Deiters (located in the bulb). This tract controls postural tone and is responsible of the maintenance of balance by adjusting head, trunk, and limb movements in response to vestibular inputs.
- *Anterior rubrospinal tract*: made up of a crossed group of fibers from the red nucleus of the mesencephalon to the cervical cord where they make a synapse with motoneurons of the ventral horn; it is also part of the extrapyramidal pathways.
- *Anterior reticular spinal tract*: a direct path from the mesencephalic and pontine reticular nuclei to the ventral horn, where motoneurons undergo a facilitatory effect.
- *Anterior (direct) cortical–spinal tract*: part of the pyramidal pathway. Fibers have their origin in the motor cortex. They pass in the internal capsule, the brain stem, and reach the anterior column. Groups of fibers cross the midline through the anterior commissure and make synapses with motoneurons of ventral horn. Although the tract is named "direct," the information carrying voluntary inputs is crossed.
- *Olivospinal tract*: from the inferior olive to the spinal anterior horn.
- *Anterior spinothalamic tract*: has fibers that emerge in the posterior horn, cross the midline, and ascend the brain stem until the lateroventral-posterior thalamic nucleus. This tract transmits information about tact and pressure. Once in the brain stem, it is joined by the lateral spinothalamic tract, both forming the spinal lemniscus.

2. Lateral column:
   - *Vestibular spinal lateral tract*
   - *Lateral rubrospinal tract*
   - *Lateral tectospinal tract*
   - *Lateral reticular spinal tract*
   - *Crossed cortical-spinal tract*: part of the pyramidal pathway and carries conscious motor information.
   - *Lateral spinothalamic tract* (STT): has fibers that start in the posterior horn, cross the midline, and ascend the brain stem up to the lateroventral-posterior thalamic nucleus. It transmits information about pain and thermal sensibility.
   - *Crossed ventral spinal cerebellar tract* (VSCT or Gowers's tract): made up of fibers originating in the Bechterew nucleus. After a first crossing of the midline, fibers ascend contralaterally

in the lateral column, reach the pons, mesencephalon, superior cerebellar peduncle, and finally get to the anterior lobe of the cerebellar cortex (see Section 3). Once in the cerebellum, they cross again. VSCT transmits proprioceptive unconscious information.

- *Direct dorsal spinal cerebellar tract* (DSCT or Flechsig's tract): has fibers that originate in the cells of Clark's nucleus (dorsal horn) and ascend homolaterally. In the bulb, they enter the cerebellum through the inferior cerebellar peduncle and end in the anterior lobe cerebellar cortex. This tract transmits proprioceptive unconscious information concerning the trunk.

3. Dorsal column:
   - *Goll fasciculus gracilis*: located medially in the dorsal column. It conveys fine tactile and conscious proprioceptive information.
   - *Burdach fasciculus cuneatus*: located laterally in the dorsal column. Fibers constituting this tract emerge from the upper thoracic and cervical spinal ganglia. In the bulb, they make synapses with the cells of the Burdach nucleus whose axons constitute the bulbothalamic tract (medial lemniscus). Fine tactile and conscious proprioceptive information is transmitted.

2.4.1 Organization of Motoneurons

Motoneurons innervating the axial musculature (trunk) are located medially. The group of motoneurons innervating the distal limbs muscles are located more laterally. There is also a ventrodorsal organization for motor neurons dedicated to extensor muscles versus those activating flexor muscles. Similarly, the central axons of the dorsal root ganglion neurons are arranged somatotopically in the dorsal column.

2.4.2 Spinal Cord and Reflexes

The spinal cord is a major site of reflex activity for the nervous system, and reflexes are important contributors in tremor genesis and maintenance. The *spinal reflex loop* results in stereotyped responses (contraction or relaxation) executed by somatic muscles in response to inputs from muscles, skin, or joint/tendon receptors. Five elements are involved in the reflex arch: receptor, afferent nerve fibers, nervous station, efferent nerve fiber, target structure (for instance: thermal receptor activated by heat → T ganglion cell branches → synapse on motoneuron → descending axon → somatic muscle resulting in movement of defense). The stretch reflex will be discussed in Section 8.

2.5 PERIPHERAL NERVES

The cranial nerves emerge from the encephalon, the spinal nerves from the spinal cord (see paragraph 2.4). Nerves can be classified as:

- *sensory nerves*: made up by sensory nervous fibers that transmit information from periphery to central structures;
- *motor nerves* (effectors): made up by motor fibers targeting muscles and glands;
- *mixed nerves*: containing both types of fibers. A spinal nerve is a typical example of a mixed nerve, containing a ventral motor root and a dorsal sensory root. Both are compacted to form the nervous trunk. As a mixed nerve, spinal nerves contain all the four types of fibers.

Cranial nerves innervate the head and neck. The vagal nerve (nerve X) reaches the thoracic and abdominal organs.

I	optic
II	olfactory
III	oculomotor
IV	trochlear
V	trigeminal
VI	abducens
VII	facial and intermediate
VIII	vestibulocochlear
IX	glossopharyngeal
X	vagus
XI	spinal accessory
XII	hypoglossal

2.6   NEUROMUSCULAR JUNCTION

The motor neuron axon innervates a specialized zone of the muscle membrane called the end-plate. Terminal axonal branches get close to the muscle fiber membrane where they form multiple grape-like varicosities (synaptic boutons) and lie in correspondence to the junctional folds of the muscle fiber membrane. Synaptic boutons release acetylcholine (Ach) that reaches the Ach receptor of the muscle membrane, opening the sodium channel. As a result, a depolarizing potential (called end-plate potential) occurs (see also Chapter 3). To produce an effective depolarization of the postsynaptic cell, the activation of the postsynaptic voltage-dependent Na^+ channel located outside the end-plate is required (Kandel and Siegelbaum, 1991).

2.7   MUSCLE

Skeletal muscle fibers are made of myofibrils whose functional contractive unit is the sarcomere. This latter is composed of fibrillar proteins called thin and thick filaments (actin, tropomyosin, and troponin). Grouped myosin molecules correspond to thick filaments. During contractions, motion of the thin filaments (relative to the thick ones) occurs. This phenomenon is called the filaments sliding. Contraction is induced by cell depolarization. The strength of contraction is related to the initial muscle length. Indeed, the muscle has elastic properties and acts as a spring in response to lengthening (see Chapter 1). Muscle stiffness increases during contraction. These changes modify filaments sliding speed, thus influencing contraction intensity (Ghez, 1991).

A given motor neuron and all muscle fibers innervated represent the *motor unit,* the smallest functional unit controlled by the motor system. The whole group of muscle fibers belonging to a motor unit has similar physiological and biochemical features.

Motor units can be classified as:

- *fast fatigable*: The unit contracts and relaxes rapidly, produces high force level but, for repeated stimuli, fatigues rapidly as well.
- *slow fatigable*: The unit has a longer contraction time and is highly resistant to fatigue, but is much less powerful than the fast fatigable motor units. These units are active mainly in tasks requiring sustained contractions.

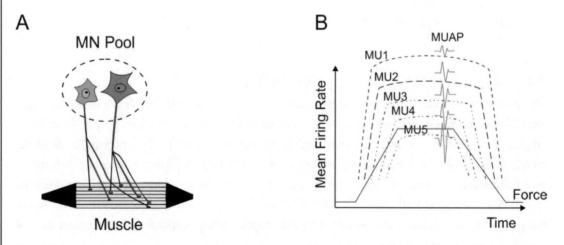

FIGURE 2.7: (a) Illustration of the innervation of a muscle fibers by two motor neurons (MN). (b) Size principle in a voluntary motor task consisting of an isometric contraction (production of a trapezoidal force). The firings of five motor units (MU1 to MU5) are illustrated. The smallest motor units are recruited first, generating the smaller motor unit action potentials (MUAPs illustrated in brown). These smaller motor units have the highest firing rates. Motor units activated subsequently have lower firing rates. Modified from Manto et al. (2008).

- *Fast fatigue resistant*: The unit has a fast fatigable-like time contraction and a slow fatigable-like resistance, but unlike the slow fatigable motor units, they can produce about twice as much force.

2.7.1   Recruitment of Motoneurons

To build up muscle contraction force, the nervous system follows the *size principle* and the *rate modulation*. Weakest inputs activate the smallest cell bodies because of their lowest threshold (Figure 2.7, see also Chapter 6 for the decomposition of the myoelectric signal). As the synaptic inputs increases in strength, larger motor neurons are progressively recruited. Increases in firing rate of motor units produce increasing force output because of the summation of successive twitches. The motor units' frequency of discharge is about 8–11 Hz during standard voluntary contraction, and it can switch up to about 25 Hz for sustained contractions. Motor units are activated asynchronously, thus minimizing jerks and producing smooth movements. According to the type of muscle fibers activated, spectral shifts in the median frequency of the power spectrum may be analyzed (see Chapter 6 about signal processing).

2.8   MUSCLE SPINDLES AND OTHER PERIPHERAL RECEPTORS

These are specialized nervous structures sensitive to a given form of energy (mechanical, thermal, chemical, etc.). The energy is transformed in electrochemical energy to enter in the sensory system to be processed. The sensory receptor is the first neuron in each sensory pathway. Localization of the sensation is elaborated, thanks to the receptive field of receptors and the central sensory neurons. Duration is encoded through the receptor adaptation to sustained stimuli, considering that receptor potential tends to decrease in response to a long duration stimulus.

The modalities of stimuli captured by the somatic sensory system are the following:

- *touch*: elicited by mechanical stimulation of body surface;
- *proprioceptive sensations*: elicited by mechanical displacements of the muscles and joints;
- *vibrations*;
- *pain*, elicited by noxious stimuli;
- *thermal sensations*, elicited by cold and warm stimuli.

Specific receptors are:

- *cutaneous and subcutaneus mechanoceptors*: Meissner's corpuscle, Pacinian corpuscle, Ruffini corpuscle, Merkel's receptor, hair-guard and hair-tylotrich, hair-down;
- *muscle spindles* (primary and secondary), Golgi tendon organs, mechanoreceptors;
- *nociceptors*: mechanical, thermal, and polymodal.

Full proprioceptive sensitivity depends on the combined actions of muscle receptors, joint receptors, and cutaneous mechanoreceptors. It is generally accepted that although cutaneous and joint receptors are involved in kinesthesia (defined as the sense of position and movement of the extremities), the main afferent signals originate from muscle spindles (Fallon and Macefield, 2007). These mechanoceptors that are entwined around specialized muscle fibers are called intrafusal fibers (Gordon and Ghez, 1991). They respond to changes in muscle length. Moreover, they discharge at higher rates in response to vibrations, which cause length variation and give the illusion of limb motion. Indeed, the illusion of joint movement is triggered by vibration (usually around 70–90 Hz) of either the muscle belly or preferably the muscle tendon. The vibration may generate a tonic vibration reflex (TVR) during which the muscle contracts involuntarily. Spindles can detect very small length changes.

Muscle spindles are innervated by groups I and II afferent fibers. Sensory afferents terminate in the central region of intrafusal fibers, while the gamma fibers innervate the polar regions relatively rich in contractile elements. Thereby, activation of a gamma efferent causes shortening of the intrafusal polar regions, which in turn, stretches the noncontractile central region. Such a stretch from both ends leads to an increase in firing rate of the sensory endings. Moreover, to grant a flow of information despite the muscle contraction (thereby fibers are shortening), the central nervous system keeps gamma motor neurons excited during the muscle contraction.

From a physiological point of view, the stretch reflex (or myotatic reflex) is the muscle contraction in response to muscle lengthening. It is linked to the reciprocal innervation that allows relaxation of the antagonist muscles. Stretch reflex consists of two components: a brisk and short-lasting phasic contraction, which is trigged by the dynamic change in muscle length, and a weaker but longer-lasting tonic contraction determined by the static stretch of the muscles. A balance exists between facilitating and inhibiting control on reflex loops. Actually, cerebral cortex and other higher centers adjust stretch reflex intensity according to the motor tasks' performance.

The activity of muscle spindles can be assessed using a technique called microneurography (Burke et al., 1978). Insulated microelectrodes (usually in tungsten) are inserted through the skin, for instance, at the level of the fibula head (knee) to record the electrical activities within the common peroneal nerve.

REFERENCES

Andersen BB, Korbo L, Pakkenberg B. A quantitative study of the human cerebellum with unbiased stereological techniques. *J Comp Neurol.* 1992;326(4):549–560. doi:10.1002/cne.903260405

Buisseret-Delmas C. Sagittal organization of the olivocerebellonuclear pathway in the rat. I. Connections with the nucleus fastigii and the nucleus vestibularis lateralis. *Neurosci Res.* 1988;5(6):475–493. doi:10.1016/0168-0102(88)90038-7

Buisseret-Delmas C, Angaut P. The cerebellar nucleocortical projections in the rat. A retrograde labelling study using horseradish peroxidase combined to a lectin. *Neurosci Lett.* 1988;84(3):255–260. doi:10.1016/0304-3940(88)90516-2

Burke D, Hagbarth KE, Lofstedt L. Muscle spindle activity in man during shortening and lengthening contractions. *J Physiol (Lond)*. 1978;277:131–142.

Carpenter MB. *Core Text of Neuroanatomy*, 3rd edn. Baltimore, MD: Williams & Wilkins, 1985.

Clendenin M, Ekerot CF, Oscarsson O. The lateral reticular nucleus in the cat. III. Organization of component activated from ipsilateral forelimb tract. *Exp Brain Res.* 1974;21(5):501–513. doi:10.1007/BF00237168

Clendenin M, Ekerot CF, Oscarsson O. The lateral reticular nucleus in the cat. IV. Activation from dorsal funiculus and trigeminal afferents. *Exp Brain Res.* 1975;24(2):131–144. doi:10.1007/BF00234059

Colin F, Ris L, Godaux E. Neuroanatomy of the cerebellum. In: *The Cerebellum and Its Disorders.* Manto MU and Pandolfo M (eds). Cambridge, UK: Cambridge University Press, 2002, pp. 6–29.

Dow RS. The evolution and anatomy of the cerebellum. *Biol Rev.* 1942;17:179–220. doi:10.1111/j.1469-185X.1942.tb00437.x

Eccles JC, Ito M, Szentagothai J. *The Cerebellum as a Neuronal Machine*. New York: Springer-Verlag, 1967.

Ekerot CF. The lateral reticular nucleus in the cat. VI. Excitatory and inhibitory afferent paths. *Exp Brain Res.* 1990;79(1):109–119.

Ekerot CF. The lateral reticular nucleus in the cat. VII. Excitatory and inhibitory projection from the ipsilateral forelimb tract (iF tract). *Exp Brain Res.* 1990;79(1):120–128.

Ekerot CF. The lateral reticular nucleus in the cat. VIII. Excitatory and inhibitory projection from the bilateral ventral flexor reflex tract (bVFRT). *Exp Brain Res.* 1990;79(1):129–137.

Fallon JB, Macefield VG. Vibration sensitivity of human muscle spindles and Golgi tendon organs. *Muscle Nerve.* 2007;36:21–29. doi:10.1002/mus.20796

Ghez C. The control of movement. In: *Principles of Neural Science.* Kandel ER, Schwartz JH, Jessel TM (eds). New York: Elsevier, 1991, pp. 530–547.

Goldman MS, Kelly PJ. The surgical treatment of tremor disorders. In: *Handbook of Tremor Disorders.* Findley LJ, Koller WC (eds). New York: Marcel Dekker, 1995, pp. 521–562.

Gorgon J and Ghez C. Muscle receptor and spinal reflex: the stretch reflex. In: *Principles of Neural Science.* Kandel ER, Schwartz JH, Jessel TM (eds). New York: Elsevier, 1991, pp. 564–580.

Ito M and Yoschida M. The cerebellar-evoked monosynaptic inhibition in Deiters' neurones. *Experientia.* 1964;20:515–516. doi:10.1007/BF02154085

Ito M. *The Cerebellum and Neural Control.* New York: Raven Press, 1984.

Kandel ER, Schwartz JH, Jessel TM. *Principles of Neural Science.* New York: Elsevier, 1991.

Kandel ER and Siegelbaum SA. Directly gated transmission of the nerve–muscle synapse. In: *Principles of Neural Science*. fKandel ER, Schwartz JH, Jessel TM (eds). New York: Elsevier, 1991, pp. 135–152.

Lainé J, Axelrad H. Lugaro cells target basket and stellate cells in the cerebellar cortex. *NeuroReport*. 1998;9(10):2399–2403. doi:10.1097/00001756-199807130-00045

Loeb C, Favale E. *Neurologia di Fazio Loeb*. Rome: Società Editirice Universo. 2003.

Manto M, Sauvage C, Roark RM. Unifying hypothesis for the motoneuronal code in neurological disorders. *Biosci Hypoth*. 2008;2:93–99. doi:10.1016/j.bihy.2008.02.011

Manto MU, Pandolfo M. *The Cerebellum and Its Disorders*. Cambridge, UK: Cambridge University Press, 2002.

Oscarsson O. Functional organization of spinocerebellar paths. In: *Handbook of Sensory Physiology, vol II. Somatosensory System*. Iggo A (ed). Berlin: Springer-Verlag, 1973, pp. 339–380.

Pouzat C, Hestrin S. Developmental regulation of basket/stellate cell → Purkinje cell synapses in the cerebellum. *J Neurosci*. 1997;17(23):9104–9112.

Serapide MF, Cicirata F, Sotelo C, Pantó MR, Parenti R. The pontocerebellar projection: longitudinal zonal distribution of fibers from discrete regions of the pontine nuclei to vermal and parafloccular cortices in the rat. *Brain Res*. 1994;644(1):175–180. doi:10.1016/0006-8993(94)90362-X

Trott JR, Apps R, Armstrong DM. Zonal organization of cortico-nuclear and nucleo-cortical projections of the paramedian lobule of the cat cerebellum. 1. The C1 zone. *Exp Brain Res*. 1998;118(3):298–315. doi:10.1007/s002210050285

CHAPTER 3

Physiology of the Nervous System

3.1 MEMBRANE POTENTIAL: BASIC CONCEPTS

Neuronal membrane acts as a barrier for the diffusion of ions, maintaining actively concentration gradients. Clouds of negative and positive ions spread over the inner and outer surfaces of the membrane. Concentration of sodium is higher in the extracellular side, and concentration of potassium is higher within the cell. These gradients are maintained thanks to the sodium–potassium pump ($Na^+K^+ATPase$). This enzyme pumps three Na^+ ions out of the cell and two K^+ ions into the cell for one ATP molecule. Moreover, neuronal cell membrane presents multiple ion channels (see Figure 3.1 for an example).

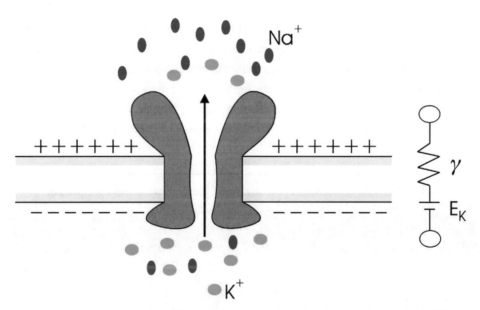

FIGURE 3.1: Representation of a cellular membrane. A specialized protein causes the efflux of potassium from the internal side of the cell (green circles). This protein is selectively permeable to this ion. A conductor γ represents the conductance of the K^+ channel. E_K represents a battery, corresponding to the potassium Nernst potential.

There is discordance among ions' diffusion down their concentration gradient, through non-gated channels, and the immobility of nonpermeable big anions that are left behind the membrane. Such a disposition determines an electrical potential difference called the *resting membrane potential* whose value in neurons is −60 to −70 mV. The electrical potential across the neuron's membrane can be calculated from the ion concentrations using the Nernst equation:

$$E = RT \times [ln(X_o/X_i)]/nF,$$

where E is the potential across the membrane, R is the resistance, T is the absolute temperature (ions move faster when the temperature increases), X_o is the concentration of the ion outside the membrane, X_i is the concentration of the ion inside the membrane, n is the charge of the ion, and F is the Faraday unit.

The equilibrium potentials of four main ions are given in Table 3.1.

The membrane potential (V_m) at any time can be estimated if the relative permeabilities to ions are known, applying the following Goldman–Hodgkin–Katz equation:

$$V_m = 2.303 \, (RT/F) \log(P_K \, [K^+]_o + P_{Na} \, [Na^+]_o + P_{Cl}[Cl^-]_i)/(P_K \, [K^+]_i \, P_{Na} \, [Na^+]_i P_{Cl} \, [Cl^-]_o)$$

Electrical signals result from brief changes of the resting membrane potential. A flux of sodium entering in (followed by potassium out) of the cell changes the potential across the membrane from −70 mV to positive value in the range of +10/+30 mV. This *depolarization* reflects the reduction of the charge separation, while *hyperpolarization* is an increase in charge separation. In response to a depolarization sufficient to reach a given threshold, the *action potential* occurs in a very short period of time (a few ms), according to the all-or-none law. By contrast, a passive electrochemical process leads to hyperpolarization, which is the electrical basis of the inhibitory response. The speed

TABLE 3.1: Equilibrium potentials calculated from the Nernst equation

ION	CONCENTRATION X_O (mM)	CONCENTRATION X_I (mM)	EQUILIBRIUM POTENTIAL (mV)
Na^+	150	15	+61.5
K^+	5.5	100	−88.2
Cl^-	125	9	−70.4
Ca^{2+}	100	10^{-4}	+183

of action potential propagation is usually directly related to the size of the axon and the presence of myelin around the axon. Myelinated axons conduct impulses faster than unmyelinated axons. Impulses typically move along neurons at a velocity from 1 to 120 m/s. The action potential is followed by a refractory period, during which a second stimulus cannot produce a second action potential.

3.2   THE SYNAPSES

The synapse (Figure 3.2) can be defined as a specialized region of membranes through which neurons are connected. There are two elemental events at the level of synapses: the electrical and the chemical transmission. In the central nervous system, chemical synapses are more abundant, and thanks to their plasticity, they represent a crucial element for processes such as perception, motion, feelings, and learning.

The chemical synapse is characterized by

- a presynaptic terminal enriched in synaptic vesicles mainly deriving from the Golgi apparatus (the end of the axon is referred to as a synaptic knob);
- a synaptic cleft of 20–40 nm (more than adjacent extracellular space);
- a postsynaptic membrane with transmembrane proteins acting as receptors and ion channels.

In response to a presynaptic action potential, vesicles release neurotransmitters. Calcium plays a key role in this process. The neurotransmitters bind subsequently to postsynaptic receptors, inducing the sodium channels' opening and generating depolarization with a postsynaptic action potential. The synapse is equipped with transporters to allow recycling of neurotransmitters. By comparison with the electrical synapses, this organization determines a delay varying from 0.3 ms up to several milliseconds. Because of the release of vesicles and neurotransmitters, chemical synapses tend to amplify nervous signals.

3.2.1   Neurotransmitters

Neurotransmitters are usually divided into excitatory and inhibitory. Their function depends on their target receptors. Each neurotransmitter works with a given specific family of receptors, maintaining a given functional role.

Typical neurotransmitters are glutamate, glycine, and GABA, which bind to ionophoric receptors undergoing a conformational change directly opening the channel. Norepinephrine and serotonin bind to a receptor that is separated from the channel and communicates with it through GTP-binding proteins and other intracellular second messengers.

In terms of synaptic delay, ionophoric receptors are faster. Transmission involving receptors activating second-messenger cascade (metabotropic receptors) can modulate the excitability of neurons and connection strength. They are implicated in learning processes, for instance.

Neurons receive both excitatory and inhibitory synapses from a huge number of axons. Therefore, nervous system neurons must *integrate* synaptic information. Processes of temporal and spatial summation occur (see below). To generate a postsynaptic action potential, a sum of several excitatory postsynaptic potentials (EPSP) should occur, and the result should be strong enough to contrast eventually inhibitory postsynaptic potentials (IPSP) generated by inhibitory synapses. The resulting synaptic inhibition is an important mechanism to control spontaneously active nerve cells, thus determining their firing pattern.

Ionic current can also be transferred directly from one neuron to another through highly specialized synapses called gap junctions. In this case, the pre- and postsynaptic cell membranes are very close (~3.5 nm). Because of this peculiar connection between neurons, an action potential in the presynaptic neuron can generate an action potential in the postsynaptic neuron. Therefore, neurons can synchronize their activity.

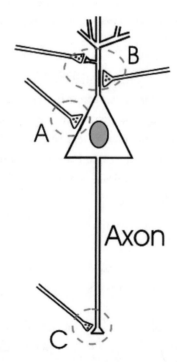

FIGURE 3.2: Categories of synapses. A: Axosomatic; B: axodendritic (left: spine synapse; right: shaft synapse); C: axo-axonic.

TABLE 3.2: Neurotransmitters in the nervous system				
MOLECULE	EFFECT	RECEPTORS	ION INVOLVED/ SECOND MESSENGERS	CENTRAL OR PERIPHERAL LOCATION
Glutamate	Excitatory	NMDA AMPA Kainate QuisqualateB	Ca^{2+}, Na^+, K^+ Na^+, K^+ Na^+, K^+ Metabotropic	Central nervous system (CNS)
GABA	Inhibitory	GABA (A) subunits α, β, γ, δ GABA (B) GABA (C)	Cl^- Metabotropic	CNS Retina and CNS
Glycine	Inhibitory	GlyR NMDA	Cl^-	CNS
Amine transmitters · Dopamine · Noradrenaline · Adrenaline · Serotonin · Histamine	Heterogeneous activities	D(1)/D(5) Adrenoceptors Adrenoceptors 5-HT1/5-HT7 H1/H4		Heterogeneously expressed in different organs and brain
Acetylcholine	Excitatory	Nicotinic AChR Muscarinic AChR	Na^+, K^+, Cl^- Metabotropic	Neuromuscular junction Autonomic nervous system CNS
Substance P and other neuropeptides		Tachykinin receptors		

3.3   EXCITATORY AND INHIBITORY SYNAPSES: NEUROTRANSMITTERS AND THEIR RECEPTORS

As mentioned earlier, postsynaptic potentials are induced by ion channels opening. Neurotransmitters (Tables 3.2 and Figure 3.3) control the permeability of ion channels.

3.3.1   Glutamate

Glutamate is the main excitatory neurotransmitter. Its action is mainly mediated by two kinds of receptors: AMPA receptor and NMDA receptor. With the first one, glutamate acts directly on a low-conductance cation channel, permeable to Na^+ and K^+. The NMDA receptor and its channel

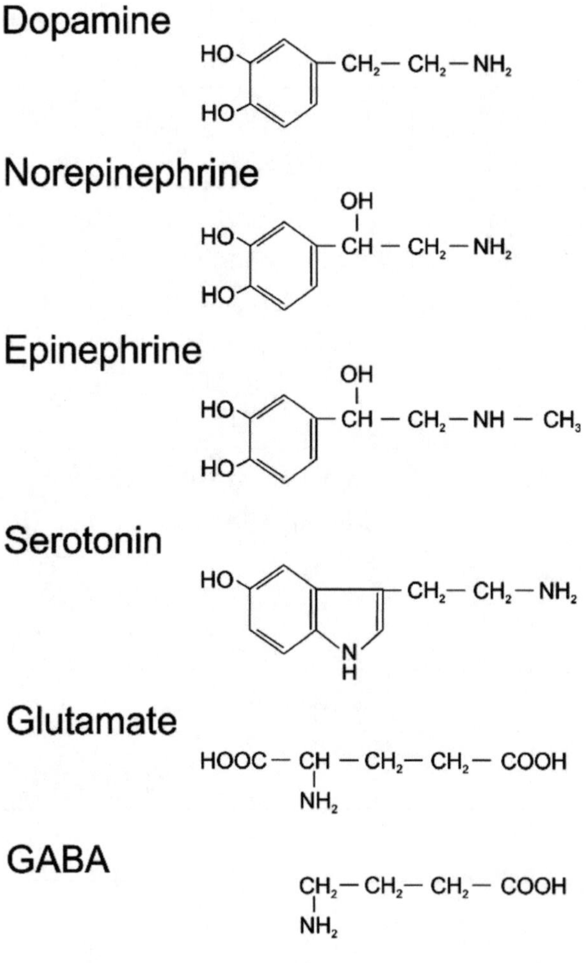

FIGURE 3.3: Main neurotransmitters of the central nervous system involved in tremor pathogenesis.

are characterized by (a) a high-conductance to Na^+, K^+, and Ca^{2+}, (b) the presence of extracellular Mg^{2+}, which delays the ionic conduction unless membrane depolarization is large enough, (c) the fact that glycine is requested for glutamate effectiveness in channel opening. Because of the Mg^{2+}-mediated block of the NMDA receptor at the normal resting membrane potential, non-NMDA receptors contribute largely in excitatory postsynaptic potential, while NMDA receptors are crucial for the late phase of EPSP. The complex NMDA channel receptor could be considered both as a voltage and a transmitter-gated channel. Moreover, NMDA-mediated Ca^{2+} entry in the cell induces a second-messenger cascade. In addition, glutamate is also known for its excitotoxicity because of free radical production in response to increasing concentration of intracellular Ca^{2+} and consecutive Ca^{2+}-dependent proteases.

3.3.2 Gamma-Aminobutyric Acid and Glycine

Gamma-aminobutyric acid (GABA) and glycine are the main inhibitory neurotransmitters. They bind directly to a channel receptor. In addition, GABA receptor has two other binding sites: one for the benzodiazepines and one for the barbiturates. The activation of these sites increases the GABA-induced Cl^- current. Moreover, the presence of any one of the three ligands influences the binding of the other two, with a global potentiation of the effect.

3.4   INTEGRATIVE ACTION OF THE NERVOUS SYSTEM

The sum of neuronal inputs—excitatory and/or inhibitory—leads to effective postsynaptic responses. According to the localization and amplitude, possible synergistic or antagonistic actions, strength of synapses, and others, postsynaptic neurons and networks integrate the inputs received and choose a corresponding behavior through the so-called integrative action, which takes place in the "axon hillock," a region of the membrane located in the initial segment of the axon with a lower threshold. When a postsynaptic potential is generated, it flows along the membrane to the hillock according to the passive properties of the membrane, mainly the time constant (summation of consecutive inputs in the same site) and the length constant (summation of postsynaptic potential that occurred in different sites). Longer time and larger space constantly increase the probability of input summation; thereby, the action potential threshold is more likely reached.

Localization is an important factor in the process of integration. Inputs are stronger when they are closer to the postsynaptic trigger zone. For instance, axodendritic synapses are excitatory, while axosomatic synapses are inhibitory. Such a disposition allows the IPSP-induced depolarization, located between the site of excitatory postsynaptic potential and the trigger zone, to prevent the excitatory input arrival. In terms of synaptic learning, this phenomenon is associated with modifications of strengths of connections between neurons. The efficacy of a synapse can change with experience, providing both memory and learning through long-term potentiation (LTP).

3.5 STRUCTURES INVOLVED IN INFORMATION PROCESSING AND MOVEMENT CONTROL

Some of the anatomical structures described in Chapter 2 have been largely investigated because they are involved in the pathogenesis of physiological and pathological tremor.

3.5.1 Muscle Spindles

Studying the behavior of the vibration-induced illusion of movement, Frima and Grünewald (2007) found out a decrease of this phenomenon with aging in healthy subjects and an impairment of this process in patients affected by essential tremor or idiopathic focal dystonia (see Chapter 5). Besides the hypothesis of abnormal central processing of the afferent information from muscle spindles, the idea that spindles themselves are involved in tremor genesis is emerging. These authors suggest that a sluggish response of the relatively inelastic muscle spindles to dynamic stimuli may result in a slightly delayed and inappropriately prolonged activation of Ia afferent fibers during muscle stretch, *adding a delay in the loop*. Moreover, in the case of essential tremor, inadequate function of the cerebellum (whose function might be to compensate by adjusting timing/gain of the feedback loop to minimise tremor) occurs. Other studies have demonstrated the systematic interaction between imposed joint movement and wrist flexor muscle activity, showing that the imposed movement of large amplitude takes progressive control over the afferent input and markedly increases the entrainment of the muscle activity via the stretch reflex (Cathers et al., 2006).

3.5.2 Spinal Cord

The spinal cord is a major site of reflex activity, and its numerous circuitries have an important role in motor coordination. The majority of reflex pathways are *polysynaptic*, composed of sensory neurons, interneurons, and motoneurons. Interneurons represent an intermediate step between afferences and efferences. Interneurons are under the supervision of supraspinal centers. Moreover, interneuron networks produce complex motor activities continuing after the end of the stimulus, thanks to *reverberating loops*. Interneurons Ia are involved in the coordination between agonist and antagonist muscles, both in reflex and voluntary movement. Renshaw inhibitory interneurons are involved in the recurrent inhibition (see Chapter 2). Inhibitory interneurons are implicated in the control of fine movements such as tactile manipulation of objects.

3.5.3 Brainstem

The brainstem contains important nuclei such as the reticular formation, vestibular nuclei, and the inferior olivary complex (see below). The brainstem nuclei integrate visual and vestibular information and regulate the segmental networks of the spinal cord.

3.5.4   Cerebellum and Inferior Olivary Complex

The cerebellum receives massive information from the periphery (see Chapter 2). In particular, medial and paramedial zones are constantly informed of the status of peripheral receptors. The cerebellum is a key piece for sensory information processing and planning of activities. It governs timing processes, meaning that it holds an internal representation of time, it controls accuracy of movements, coordination, and it is involved in learning. In particular, it tunes dynamic control of learned movement (Sánchez-Campusano et al., 2007). The cerebellum is critically involved in predictions for fast movements and online regulation of slow movements (Diedrichsen et al., 2007).

The mossy fiber–granule cell system carries more temporal information than the climbing fiber system.

3.5.5   Basal Ganglia

The basal ganglia influence movement via thalamocortical projections and via projections to the brainstem nuclei (Pong et al., 2007). The internal segment of the globus pallidus (GPi) receives motor-related cortical signals mainly through the striatum, the external segment of the globus pallidus (GPe), and the subthalamic nucleus (STN). The GPi sends its outputs outside the basal ganglia and plays a key role in motor control (Tachibana et al., 2008).

Disconnecting the basal ganglia motor pathway consistently produces bradykinesia and hypometria, but seldom affects movement initiation time, movement guidance, or the capacity to produce iterative reaches (Desmurget and Turner, 2007). Temporal preparation is disrupted in case of basal ganglia dysfunction (Praamstra and Pöpe, 2007).

3.5.6   Motor Cortex

Neurons in the motor cortices have uniformly distributed preferred directions of activity, consistent with general purpose directional calculations. There is evidence for an orderly mapping of the preferred direction of voluntary motion in the motor cortex (Georgopoulos et al., 2007).

3.5.7   Premotor Cortex

The upper motor neurons in the premotor cortex influence motor behavior through reciprocal connections with the primary motor cortex. One of the core brain areas involved in visually guided reaching is the dorsal aspect of the premotor cortex (PMC; Batista et al., 2007). Lateral premotor cortex is involved in the selection of a movement or a sequence of movements dependent of external cues. Medial parts of the premotor cortex mediate movement based on internal cues. Premotor cortex exerts a control on proximal muscles.

3.5.8   Posterior Parietal Cortex

The posterior parietal cortex (PPC) is responsible for transforming visual information into motor commands. It processes spatial information and the control of eye movements. It contributes to spatial attention. PPC damage after a stroke often leads to the clinical syndrome of neglect, in which patients seem unable to focus attention on events in the contralesional hemifield.

REFERENCES

Batista AP, Santhanam G, Yu BM, Ryu SI, Afshar A, Shenoy KV. Reference frames for reach planning in macaque dorsal premotor cortex. *J Neurophysiol.* 2007;98(2):966–983. doi:10.1152/jn.00421.2006

Cathers I, O'Dwyer N, Neilson P. Entrainment to extinction of physiological tremor by spindle afferent input. *Exp Brain Res.* 2006;171(2):194–203. doi:10.1007/s00221-005-0258-9

Desmurget M, Turner RS. Testing basal ganglia motor functions through reversible inactivations in the posterior internal globus pallidus. *J Neurophysiol.* 2008;99(3):1057–1076. doi:10.1152/jn.01010.2007

Diedrichsen J, Criscimagna-Hemminger SE, Shadmehr R. Dissociating timing and coordination as functions of the cerebellum. *J. Neurosci.* 2007;27(23):6291–6301. doi:10.1523/JNEUROSCI.0061-07.2007

Frima N, Grünewald RA. Abnormal vibration induced illusion of movement in essential tremor: evidence for abnormal muscle spindle afferent function. *J Neurol Neurosurg Psychiatry.* 2005;76(1):55–57. doi:10.1136/jnnp.2004.036640

Georgopoulos AP, Merchant H, Naselaris T, Amirikian B. Mapping of the preferred direction in the motor cortex. *Proc Natl Acad Sci USA.* 2007;104(26):11068–11072. doi:10.1073/pnas.0611597104

Kandel ER, Schwartz JH, Jessel TM. *Principles of Neural Science.* New York: Elsevier 2001.

Marshall FH. The role of GABA(B) receptors in the regulation of excitatory neurotransmission. *Results Probl Cell Differ.* 2008;44:87–98.

Pong M, Horn KM, Gibson AR. Pathways for control of face and neck musculature by the basal ganglia and cerebellum. *Brain Res Rev.* 2007. doi:10.1016/j.brainresrev.2007.11.006

Praamstra P, Pope P. Slow brain potential and oscillatory EEG manifestations of impaired temporal preparation in Parkinson's disease. *J Neurophysiol.* 2007;98(5):2848–2857. doi:10.1152/jn.00224.2007

Sánchez-Campusano R, Gruart A, Delgado-García JM. The cerebellar interpositus nucleus and the dynamic control of learned motor responses. *J Neurosci.* 2007;27(25):6620–6632. doi:10.1523/JNEUROSCI.0488-07.2007

Schmidt M. GABA(C) receptors in retina and brain. *Results Probl Cell Differ.* 2008;44:49–67.

Tachibana Y, Kita H, Chiken S, Takada M, Nambu A. Motor cortical control of internal pallidal activity through glutamatergic and GABAergic inputs in awake monkeys. *Eur J Neurosci.* 2008;27(1):238–253.

.   .   .   .

CHAPTER 4

Characterization of Tremor

From the clinical standpoint, tremor is classically classified as rest tremor, postural tremor, and kinetic tremor. Each category of tremor has its own characteristics in terms of:

- frequency
- amplitude
- body segments involved
- distribution (symmetry)
- enhancing/reducing effect of tasks or positions
- associated disabilities.

Neurological examination of a patient complaining and/or presenting tremor includes the following tests or maneuvers (see Chapter 6 for the clinical evaluation of tremor):

- evaluation of oculomotor movements;
- the patient is asked to repeat standard sentences to evaluate dysarthria or other speech deficits;
- the presence of vertical or side-to-side head movements is noted;
- assessment of drawing Archimedes' spiral (see Chapter 6);
- concomitant abnormal movements (included dystonia, tics) are looked for;
- the observation of abnormal movements eventually occurring at rest, the occurrence of the tremor during postural task, or voluntary muscle contraction;
- a possible amyotrophy or muscle hypertrophy is looked for;
- detailed sensory testing (light touch, sense of position, vibratory testing, pin-prick threshold, hot and cold stimuli);
- evaluation of tendon/plantar reflexes;
- the following clinical maneuvers are assessed:
  - holding the upper limbs outstretched parallel to the floor ("Barré test");
  - finger-to-nose test, index-to-index test, knee–tibia test and "Mingazzini test";

TABLE 4.1: Main disorders associated with tremor	
TYPE OF TREMOR	**DISEASES**
Rest tremor	Parkinson's disease Drug-induced Parkinsonism Stroke Posttraumatic tremor
Postural tremor	Essential tremor Enhanced physiological tremor Cerebellar diseases Multiple sclerosis Posttraumatic tremor Metabolic diseases
Kinetic tremor	Cerebellar diseases Essential tremor Multiple sclerosis

- ° entrainment test (see Section 4.6);
- ° Romberg position test;
- ° regular and tandem gait.

Additional clinical tests, metrical scales, laboratory assessment and others tools for quantification of tremor will be described in detail in Chapter 6.

The main disorders associated with postural, rest, and kinetic tremor are given in Table 4.1.

4.1   PHYSIOLOGICAL TREMOR AND ENHANCED PHYSIOLOGICAL TREMOR

Healthy subjects present a physiological tremor (see also Chapter 1), an involuntary rhythmical movement of limb segments typically in the frequency range of 8–12 Hz. The amplitude is small and is barely seen with the naked eye (Cathers et al., 2006). Physiological tremor's frequency is reduced by addition of inertia (Elble, 2003; see Figure 1.1a) and is age-independent (Raethjen et al., 2000). The maintenance of a postural position or action has a typical pronouncing effect on physiological tremor during a standing position, leading to an increase of the amplitude. This is more obvious in the elderly persons (Morrison et al., 2006).

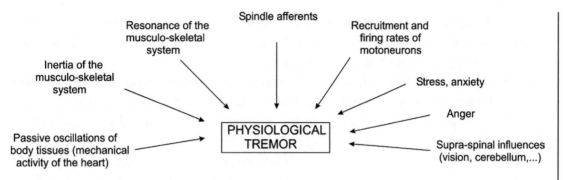

FIGURE 4.1: Factors influencing physiological tremor.

TABLE 4.2: Drugs enhancing physiological tremor	
Amiodarone	Neuroleptics
Antihistamine drugs (terfenadine)	Nicotine
β2-Adrenergic agonists (bronchodilators)	Nifedipine
Cinnarizine	Phenelzine
Cyclosporin A	Phenylpropanolamine
Diazepam withdrawal	Pindolol
Donezepil	Prednisone
Flunarizine	Procainamide
Fluoxetine	Pseudoephedrine
Lamotrigine	Theophylline
Lithium	Thyroxine
Methylphenidate	Tricyclic antidepressants
Metoclopramide	Valproic acid
Monoamine oxidase inhibitors	

From Bhidayasiri (2005), Benito-León and Louis (2006), and Song et al. (1993, 2008).

Physiological tremor is augmented by hypoxia (Krause et al., 2000). Mental stress increases its amplitude (this is not specific) but decreases its frequency (Growdon et al., 2000). Factors influencing physiological tremor are summarized in Figure 4.1.

Several metabolic conditions (mainly thyrotoxicosis or hypoglycaemia) and drugs (see Table 4.2) can enhance physiological tremor giving the so-called enhanced physiological tremor, a visible high frequency postural tremor. Muscle fatigue may trigger or exaggerate this tremor. Caffeine significantly increases whole-arm physiological tremor (Miller et al., 1998), while ethanol usually decreases its magnitude (Lakie et al., 1994). Enhancement of postural tremor in subjects treated with amitryptiline (a tricyclic antidepressant drug) is a common phenomenon. Intravenous infusion of adrenaline or noradrenaline increases physiological tremor. This is antagonized by administration of beta-adrenoreceptor blockers (Marsden et al., 1967). Frequency, amplitude, burst duration, and burst pattern of drug-induced tremors can mimic essential tremor (Mehndiratta et al., 2005).

4.2   REST TREMOR

Rest tremor occurs while the body segment is maintained at rest. Rest tremor is typically in the 3- to 6-Hz frequency range. During a clinical evaluation, rest tremor is assessed with the patient sitting with his arms supported against gravity, without any voluntary activities. Tremor may disappear with action. For this reason, rest tremor is often a cause of social embarrassment, rather than disability. Rest tremor is usually asymmetrical, starting distally in the arms. Lips and jaw can be affected. Typically, tremor in the upper limbs reminds the "pill-rolling" movement. In some cases, patients can reduce the tremor by holding one hand with the other or crossing the legs. Rest tremor may increase with mental stress (i.e., counting backwards) or contralateral motion. This is not specific to rest tremor.

The most common cause of rest tremor is idiopathic Parkinson's disease (PD), which is detailed in Chapter 5.

4.3   POSTURAL TREMOR

Postural tremor occurs in body parts during the maintenance of a posture, such as holding a cup. This form of tremor is triggered by maintaining a position against gravity. Some authors consider that physiological tremor and enhanced physiological tremor are a peculiar form of postural tremor. Postural tremor often causes a significant disability. The frequency of postural tremor is usually between 4 and 12 Hz.

Many disorders are associated with postural tremor. Essential tremor (ET) is a typical example (Figure 4.2, see also Chapter 5). Lesions in different anatomical locations of the cerebellar

ESSENTIAL TREMOR

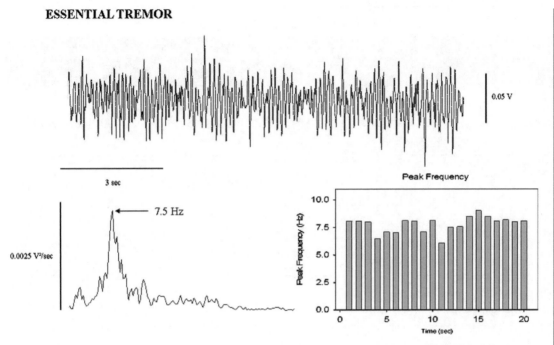

FIGURE 4.2: Monoaxial accelerometry in essential tremor. The patient is asked to maintain the upper limbs horizontally parallel to the floor. A peak frequency of 7.5 Hz is clearly identified on the power spectrum.

pathways result in different kinds of postural tremor (see Table 4.3). An example of postural tremor associated with a cerebellar stroke is illustrated in Figure 4.3. Acute or subacute postural tremor may result from cortical lesions such as mass occupying lesions, ischemic lesions, and arteriovenous malformations (Balci et al., 2007). Postural tremor of the upper limbs is a common manifestation of Wilson's disease, very likely generated within a synchronized cerebellothalamocortical network, comprising the primary sensory–motor cortex, higher cortical motor areas (supplementary motor area, premotor cortex), and posterior parietal cortex (Südmeyer et al., 2006).

4.4   KINETIC TREMOR

Tremor appearing during limb movement and often worsening near the target is defined as a kinetic tremor. Tremorous movements are perpendicular to the main direction of the intended movement and tend to be predominant over proximal musculature. The terms "intention tremor" and "terminal tremor" have also been used in the past. Kinetic tremor is classically tested during finger-to-nose or

		TABLE 4.3: Cerebellar postural tremors		
TREMOR TYPE	**PATHOLOGY**	**PRECIPITANTS**	**FREQUENCY (Hz)**	**DISTRIBUTION**
Midbrain tremor	Cerebellar outflow tracts Nigrostriatal lesion	Any posture	2–5	Distal > proximal
Asthenic cerebellar tremor	Cerebellar hemisphere	Fatigue/weakness	Irregular	Proximal + distal
Precision cerebellar tremor	Cerebellar nuclei and brachium conjunctivum	Accurate placements	2–5	Distal
Cerebellar axial postural tremor	Cerebello-olivary system	Any posture	2–10	Proximal > distal
Cerebellar proximal exertional tremor	Cerebellar malformation	Prolonged exercise	3–4	Proximal > distal

Adapted from Brown et al. (1997).

knee–tibia tests. Tremor occurring in cerebellar diseases is the typical example of a kinetic tremor, affecting especially the head and the upper part of the body. The frequency is between 2 and 7 Hz in the large majority of cases. In case of cerebellar injury, kinetic tremor may increase secondarily, involving especially in the proximal joints (shoulder). Addition of inertia tends to improve kinetic tremor. A recent study on the differences between smokers and nonsmokers in terms of kinetic hand tremor showed a greater tremor intensity in smokers, even more apparent in women (Louis, 2007).

FIGURE 4.3: Postural tremor in a patient presenting a cerebellar stroke in the territory of the superior cerebellar artery. (Top) Axial MRI of the posterior fossa: stroke (arrow) in the territory of the superior cerebellar artery. B: brainstem, C.H.: cerebellar hemisphere, V: vermis. The lesion involves the outflow tract of the cerebellar nuclei. (Bottom) Data from a monoaxial accelerometer and from a surface electromyographic (EMG) sensor are shown, as well as the corresponding power spectra. Note the waveform aspect, which is characterized by asymmetry.

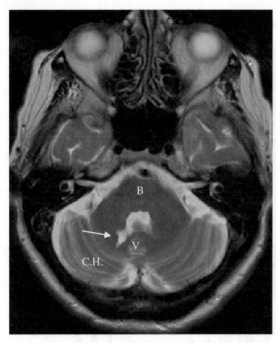
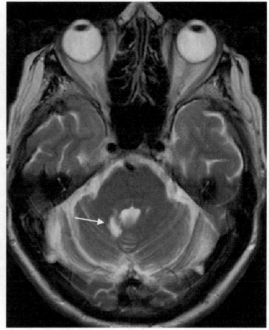

POSTURAL TREMOR—CEREBELLAR STROKE

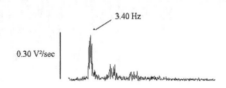

3.40 Hz

0.30 V²/sec

Accelerometer - Upper limb outstreched

10 sec

Flexor Carpi Radialis - Rectified EMG

10 sec

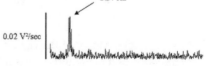

3.37 Hz

0.02 V²/sec

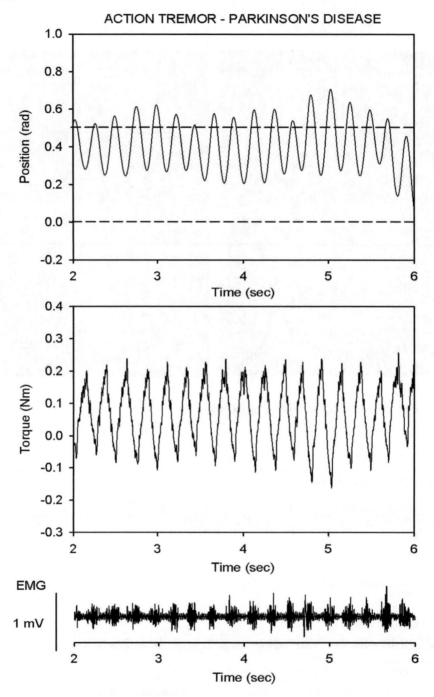

FIGURE 4.4: Action tremor of the right wrist in a patient presenting Parkinson's disease. (Top) Position. The tremor is well discernible. (Middle) Simultaneous torques. (Bottom) Surface EMG activities of the extensor carpi radialis (right side).

4.5   ACTION TREMOR

"Action tremor" refers to any tremor produced by voluntary contraction of muscles. It includes postural, isometric, and kinetic tremor.

Isometric tremor occurs when a voluntary muscle contraction is opposed by a rigid stationary object (Findley and Koller, 1995). Prolonged action-induced clonus, in the case of lesions of the central motor pathways, may mimic action tremor (Fraix et al., 2008).

4.6   OTHER FORMS OF TREMOR

4.6.1 Psychogenic Tremor

Psychogenic tremor is the most common form (55 %) of psychogenic movement disorder (Bhatia and Schneider, 2007). Psychogenic tremor is usually made of a combination of rest, postural, and

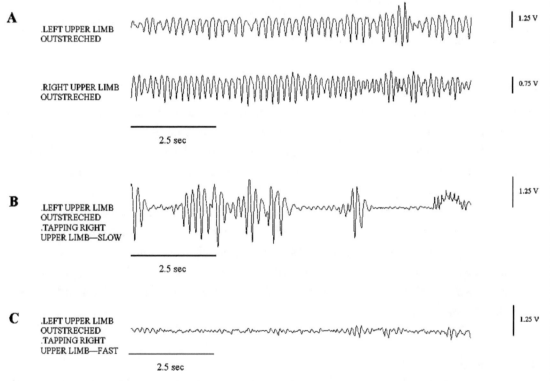

FIGURE 4.5: Psychogenic postural tremor. (a) Accelerometric measurement in both upper limbs during a postural task. Tremor appears regular bilaterally. (b) When the patient performs slow tapping movements with the right upper limb, tremor in the left upper limbs becomes highly variable in intensity and frequency. (c) During fast tapping of the right upper limb, the intensity of tremor decreases markedly on the contralateral upper limb.

kinetic components. It often involves preferentially the upper limbs, followed by the head and legs. It may be continuous or intermittent with typical fluctuations in frequency and amplitude. Psychogenic tremor is often characterized by an abrupt onset, periods of spontaneous remissions, unresponsiveness to antitremor drugs, and absence of concomitant neurological signs (Bhidaysiri, 2005). Preceding events include work-related injuries or other accidents. Psychogenic tremor is influenced by distraction and is entrained at the frequency of a rhythmic movement in the opposite limb (entrainment test), a maneuver which is helpful in daily practice to distinguish a pseudotremor. Measurement with accelerometers of the frequency changes during tapping in a group of psychogenic tremor patients (compared with ET and PD patients) showed large tremor frequency changes and a high intraindividual variability with tapping (Figure 4.5; Zeuner et al., 2003). Comorbidity with psychiatric disorders is common.

4.6.2 Task-Specific Tremor

Task-specific tremor occurs during given actions, for instance, writing representing in this case the primary writing tremor (see also Chapter 5).

4.6.3 Dystonic Tremor

Dystonic tremor is mainly a postural and kinetic tremor in an extremity or in a body part affected by dystonia. It is absent at rest in about half of the cases. Its frequency is typically irregular, varying from 4 to 9 Hz. Amplitude is unsteady. It is usually asymmetrical and remains localized. A typical example is the tremulous spasmodic torticollis. Patients affected may use the "geste antagoniste" or sensory tricks to reduce the tremor (Bhidayasiri, 2005).

Jedynak et al. (1991) studied the clinical and electromyographic characteristics of tremor in patients with idiopathic dystonia. They characterized dystonic tremor as follows: postural, localized, and irregular in amplitude and periodicity, absent during muscle relaxation, exacerbated by smooth muscle contraction, and frequently associated with myoclonus. Although it looks like ET, dystonic tremor is a distinct entity. Dystonic tremor is more irregular with a broader range of frequencies and is usually asymmetrical.

4.6.4 Holmes' Tremor

Holmes' tremor (also called midbrain tremor) is a symptomatic tremor affecting predominantly proximal segments. It has a low frequency (<4.5 Hz). Oscillations are present at rest but worsen during movement and goal-directed tasks. A delay between 4 weeks to 2 years is commonly observed between lesion onset and occurrence of tremor, suggesting a rearrangement of central pathways in the brain or an aberrant plasticity. Lesions are most often located in the upper brain stem, thalamus, and cerebellum, interrupting the pathways in the midbrain tegmentum (Guillain–Mollaret triangle, rubrospinal fibers, nigrostriatal fibers) and the serotoninergic brainstem telencephalic fibers. A unique case of reversible Holmes' tremor associated with hyperglycemia was recently described (Tan et al., 2006).

4.6.5 Cortical Tremor

Cortical tremor is an improper term that actually refers to a specific rhythmic myoclonus. This kind of tremor is characteristic of the familial cortical myoclonic tremor, a rare disorder often misdiagnosed as ET (Bourdain et al., 2006; see Chapter 5).

4.6.6 Orthostatic Tremor

Orthostatic tremor (OT) is a high-frequency tremor (13–18 Hz) predominantly in the legs and trunk (Figure 4.6). The term was first introduced in 1984 as a tremor involving mainly the legs and trunk (Heilman, 1984), although the first description was reported in 1970 (Pazzaglia et al., 1970). OT is absent when the patient walks, is sitting, or lies down. OT is triggered during isometric contraction of the limb muscles (Piboolnurak et al., 2005). OT is divided into primary orthostatic tremor (POT; no lesion detected with imaging techniques) and secondary orthostatic tremor (SOT; see Chapter 5). SOT is different from the paroxysmal high-frequency tremor occurring in up-

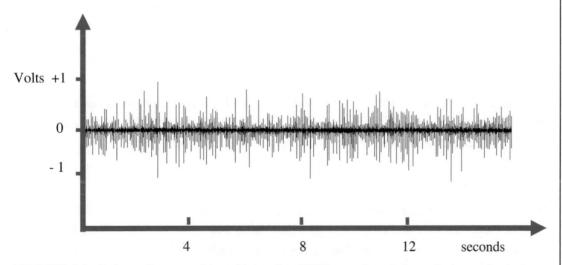

FIGURE 4.6: Orthostatic tremor detected by surface EMG recordings in lower limb muscles during a standing tasks. Amplifier gain: 1000. Frequency of discharges: 17 Hz.

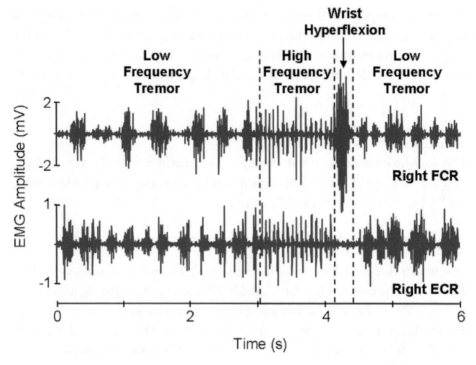

FIGURE 4.7: Episodes of bilateral high-frequency synchronous discharges (BHFSD) occurring in a patient with atrophy of the posterior fossa. The patient exhibits a low frequency tremor followed by paroxysmal synchronous bursts of electromyographic (EMG) activities in the right flexor carpi radialis (FCR) and right extensor carpi radialis (ECR), probably resulting from a sudden activation of cerebellar reverberating circuits. High-frequency bursts are terminated by a forceful wrist hyperflexion. Activities in the FCR and ECR muscles are detected using surface EMG electrodes.

per limbs in patients with posterior fossa lesions (bilateral high-frequency synchronous discharges BHFSD; Figure 4.7). BHFSD affect the upper limbs, occur in a context of low-frequency tremor (3–5 Hz), are intermittent, and are terminated by a forceful flexion of the wrist.

4.6.7 Palatal Tremor

Palatal tremor is a rhythmic involuntary movement of the soft palate (Deuschl et al., 1994) not abolished by sleep, with a range of frequency of 0.3–100 Hz (usually in the range 1.5–3 Hz). Patients often complain of clicking in the ears. Synchronous jerks may affect simultaneously the eyes, the face, the larynx, the neck, the shoulder, and the diaphragm (palatopharyngolaryngo-oculodiaphragmatic myoclonus"; Guillain, 1938). Also known as "palatal myoclonus," palatal tremor is either essential or secondary to a lesion in the brainstem or along cerebellar pathways (Guillain–Mollaret triangle). Ciprofloxacin-induced palatal tremor has been described recently (Cheung et al., 2007).

4.6.8 Tremor After a Peripheral Nerve Injury

Tremor after a peripheral nerve injury appears after a lesion of the peripheral nervous system (Costa et al., 2006). Neurogenic changes may be detected clinically or with neurophysiological investigations in the territory of the damaged nerve (nerve conduction velocities, needle EMG studies, somatosensory-evoked potentials). Concomitant dystonia is present in about 30% of the cases.

REFERENCES

Balci K, Utku U, Cobanoglu S. Two patients with tremor caused by cortical lesions. *Eur Neurol.* 2007;57(1):36–38. doi:10.1159/000097008

Benito-León J, Louis ED. Essential tremor: emerging views of a common disorder. *Nat Clin Pract Neurol.* 2006;2(12):666–678. doi:10.1038/ncpneuro0347

Bhatia KP, Schneider SA. Psychogenic tremor and related disorders. *J Neurol.* 2007;254(5):569–574. doi:10.1007/s00415-006-0348-z

Bhidayasiri R. Differential diagnosis of common tremor syndromes. *Postgrad Med J.* 2005;81:756–762. doi:10.1136/pgmj.2005.032979

Bourdain F, Apartis E, Trocello JM, Vidal JS, Masnou P, Vercueil L, Vidailhet M. Clinical analysis in familial cortical myoclonic tremor allows differential diagnosis with essential tremor. *Mov Disord.* 2006;21(5):599–608. doi:10.1002/mds.20725

Brown P, Rothwell JC, Stevens JM, Lees AJ, Marsden CD. Cerebellar axial postural tremor. *Mov Disord.* 1997;12(6):977–984. doi:10.1002/mds.870120622

Cathers I, O'Dwyer N, Neilson P. Entrainment to extinction of physiological tremor by spindle afferent input. *Exp Brain Res.* 2006;171(2):194–203. doi:10.1007/s00221-005-0258-9

Cheung YF, Wong WW, Tang KW, Chan JH, Li PC. Ciprofloxacin-induced palatal tremor. *Mov Disord.* 2007;22(7):1038–1043. doi:10.1002/mds.21452

Costa J, Henriques R, Barroso C, Ferreira J, Atalaia A, de Carvalho M. Upper limb tremor induced by peripheral nerve injury. *Neurology.* 2006;67(10):1884–1886. doi:10.1212/01.wnl.0000244437.31413.2b

Deuschl G, Toro C, Valls-Solé J, Zeffiro T, Zee DS, Hallett M. Symptomatic and essential palatal tremor. 1. Clinical, physiological and MRI analysis. *Brain.* 1994;117(Pt 4):775–788.

Elble RJ. Characteristics of physiologic tremor in young and elderly adults. *Clin Neurophysiol.* 2003;114(4):624–635. doi:10.1016/S1388-2457(03)00006-3

Findley LJ, Koller WC. *Handbook of Tremor Disorders.* New York: Marcel Dekker, 1995.

Fraix V, Delalande I, Parrache M, Derambure P, Cassim F. Action-induced clonus mimicking tremor. *Mov Disord.* 2008;23(2):285–288. doi:10.1002/mds.21826

Growdon W, Ghika J, Henderson J, van Melle G, Regli F, Bogousslavsky J, Growdon JH. Effects of proximal and distal muscles' groups contraction and mental stress on the amplitude and frequency of physiological finger tremor. An accelerometric study. *Electromyogr Clin Neurophysiol.* 2000;40(5):295–303.

Guillain G. The syndrome of synchronous and rhythmic palato-pharyngo-laryngo-oculo-diaphragmatic myoclonus. *Proc R Soc Med.* 1938;31:1031–1038.

Heilman KM. Orthostatic tremor. *Arch Neurol.* 1984;41(8):880–881.

Jedynak CP, Bonnet AM, Agid Y. Tremor and idiopathic dystonia. *Mov Disord.* 1991;6(3):230–236. doi:10.1002/mds.870060307

Krause WL, Leiter JC, Marsh Tenney S, Daubenspeck JA. Acute hypoxia activates human 8–12 Hz physiological tremor. *Respir Physiol.* 2000;123(1–2):131–141. doi:10.1016/S0034-5687(00)00146-8

Lakie M, Frymann K, Villagra F, Jakeman P. The effect of alcohol on physiological tremor. *Exp Physiol.* 1994;79(2):273–276.

Louis ED. Kinetic tremor: differences between smokers and non-smokers. *Neurotoxicology.* 2007;28(3):569–575. doi:10.1016/j.neuro.2006.12.006

Manto MU, Pandolfo M. *The Cerebellum and Its Disorders.* Cambridge, UK: Cambridge University Press, 2002.

Marsden CD, Foley TH, Owen DAL, McAllister RG. Peripheral beta-adrenergic receptors concerned with tremor. *Clin Sci.* 1967;33:53–65.

Mehndiratta MM, Satyawani M, Gupta S, Khwaja GA. Clinical and surface EMG characteristics of valproate induced tremors. *Electromyogr Clin Neurophysiol.* 2005;45(3):177–182.

Miller LS, Lombardo TW, Fowler SC. Caffeine, but not time of day, increases whole-arm physiological tremor in non-smoking moderate users. *Clin Exp Pharmacol Physiol.* 1998;25(2):131–133. doi:10.1111/j.1440-1681.1998.tb02190.x

Morrison S, Mills P, Barrett R. Differences in multiple segment tremor dynamics between young and elderly persons. *J Gerontol A Biol Sci Med Sci.* 2006;61(9):982–990.

Pazzaglia P, Sabattini L, Lugaresi E. Su di un singolare disturbo della stazione eretta (osservazione di tre casi). *Riv Frenatria.* 1970;96:450–457.

Piboolnurak P, Yu QP, Pullman SL. Clinical and neurophysiologic spectrum of orthostatic tremor: case series of 26 subjects. *Mov Disord.* 2005;20(11):1455–1461. doi:10.1002/mds.20588

Raethjen J, Pawlas F, Lindemann M, Wenzelburger R, Deuschl G. Determinants of physiologic tremor in a large normal population. *Clin Neurophysiol.* 2000;111(10):1825–1837. doi:10.1016/S1388-2457(00)00384-9

Song IU, Kim JS, Ryu SY, Lee SB, An JY, Lee KS. Donepezil-induced jaw tremor. *Parkinsonism Relat Disord.* 2008. doi:10.1016/j.parkreldis.2008.01.003

Soto J, Sacristan JA, Alsar MJ, Sainz C. Terfenadine-induced tremor. *Ann Neurol.* 1993; 33(2):226. doi:10.1002/ana.410330216

Südmeyer M, Pollok B, Hefter H, Gross J, Butz M, Wojtecki L, Timmermann L, Schnitzler A. Synchronized brain network underlying postural tremor in Wilson's disease. *Mov Disord.* 2006;21(11):1935–1940. doi:10.1002/mds.21104

Tan JH, Chan BP, Wilder-Smith EP, Ong BK. A unique case of reversible hyperglycemic Holmes' tremor. *Mov Disord.* 2006;21(5):707–709. doi:10.1002/mds.20795

Zeuner KE, Shoge RO, Goldstein SR, Dambrosia JM, Hallett M. Accelerometry to distinguish psychogenic from essential or parkinsonian tremor. *Neurology.* 2003;61(4):548–550.

·   ·   ·   ·

CHAPTER 5

Principal Disorders Associated with Tremor

In this chapter, we describe the diseases which may present with tremor as a main symptom.

5.1   ESSENTIAL TREMOR

Essential tremor (ET) is no more viewed as a monosymptomatic benign disorder, and its clinical spectrum includes both motor and nonmotor features (Benito-Leon and Louis, 2006; Thanvi et al., 2006). ET is one of the most common neurological disorders among adults and is the most common movement disorder in the elderly. Its prevalence is 4% in the population above 65 years of age, with a higher prevalence in White than in African-American populations. No gender differences are reported (Thanvi et al., 2006). A recent prospective, population-based study in Spain of individuals aged 65 years and older revealed an adjusted incidence of ET of 616 per 100,000 person-year (Benito-Leon and Louis, 2006). Incidence and prevalence of ET increase with aging, but it should be kept in mind that the disease can occur at any age. In a screening of 2227 people in Turkey, the prevalence of ET was 3.09% in subjects aged over 18 years (Sur et al., 2008). Several studies have reported an increased risk to develop Parkinson's disease in ET populations. The finding of hyper-echogenicity of the substantia nigra—considered as a transcranial sonography risk marker—has been recently proposed as a predictive sign of such a risk (Stockner et al., 2007).

ET is a familial disease in many cases. One possible genetic pattern is a dominant mode of inheritance with an incomplete penetrance by the age of 65 years (Bain et al., 1994). So far, three genetic susceptibility loci (ETM1 on 3q13, ETM2 on 2p24.1, and a locus on 6p23) have been identified in families with the disorder (Deng et al., 2007). In addition, some ET patients have the gly9 susceptibility to the DRD3 gene (dopamine D3 receptor encoding region), which might cause an increased excitability of neuronal circuits. A recent study underlined the possibility of a deficiency of mtDNA multicomplexes in ET (Yoo et al., 2008). Nevertheless, the nonfamilial sporadic form of ET is recognized as well. The familial form of ET is characterized by an earlier age of onset than the sporadic form (Louis and Ottman, 2006).

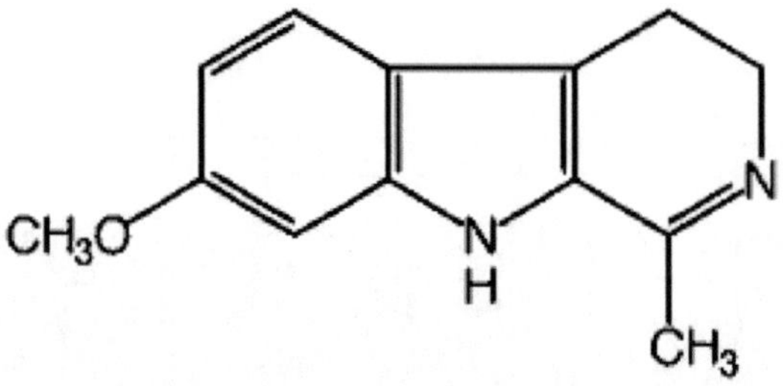

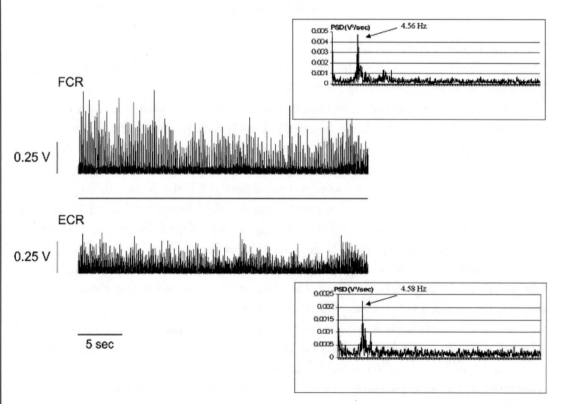

FIGURE 5.1: Structure of a beta-carboline alkaloid (BCA).

FIGURE 5.2: Postural tremor in the upper limbs during the task of maintaining the upper limbs outstretched parallel to the floor in a patient presenting essential tremor. (Left) Recordings in the right flexor carpi radialis (FCR) and extensor carpi radialis (ECR) using surface electromyography (EMG). Rectified traces are shown. (Right) Power spectra show peaks of frequency, respectively, at 4.56 and 4.58 Hz. Bursts of EMG activity are synchronous in the two antagonist muscles.

ET might be triggered by environmental factors (Jiminez-Jiminez et al., 2007). The β-carboline alkaloids (BCA, Figure 5.1) are the most studied toxic factors. These toxins are normally present in the human diet. Higher concentrations have been reported in blood of ET patients compared to control subjects (Louis et al., 2002).

ET is characterized by a bilateral postural tremor of moderate to severe amplitude at a frequency of 4–12 Hz (Figure 5.2). A kinetic tremor is often observed in association with the postural

TABLE 5.1: Clinical diagnostic criteria for definite ET
Inclusion criteria
Bilateral postural tremor with or without kinetic tremor, involving hands and forearms, that is visible and persistent
Duration >5 years
Exclusion criteria
Other abnormal neurological signs (except Froment's sign*)
Presence of known causes of increased physiological tremor
Concurrent or recent exposure to tremorogenic drugs or the presence of a drug withdrawal state
Direct or indirect trauma to the nervous system within 3 months before the onset of tremor
Historical or clinical evidences of psychogenic origins
Convincing evidence of sudden onset or evidence of stepwise deterioration
Probable ET
Tremor may be confined to one body part other than the hands, and duration is longer than 3 years
Possible ET
Other neurological disorders are present (parkinsonism) or other isolated tremors (task-specific tremor)

*Froment's sign: rhythmical resistance to passive movement of a limb about a joint that can be detected when there is a voluntary action of another body part.

oscillations (Brennan et al., 2002). A rest tremor may also be present, usually in the case of a longer disease duration (Cohen et al., 2003). A bimodal distribution of tremor frequencies is relatively common in ET.

Under the same conditions of observation, there is usually little variation in frequency for postural tremor of the hands (usually <0.6 Hz). With time, tremor amplitude tends to increase, whereas the frequency decreases (Elble et al., 1994). ET appears most frequently in the hands, followed by the head, voice, tongue, legs, and trunk. Typically, a more or less symmetrical tremor affects the extremities and progresses medially. Head tremor is milder than limb tremor and usually is of the side-to-side type. Alcohol intake temporary reduces ET in up to 75% of patients (Paulson, 1892; Lou and Jankovic, 1991). Diagnosis of ET is mainly clinical, based on medical history and physical examination. Clinical diagnostic criteria for definite, probable, or possible ET, according to the Consensus statement of the Movement Disorder Society on Tremor, are reported in Table 7 (Deuschl et al., 1998).

5.2   IDIOPATHIC PARKINSON'S DISEASE

Parkinson's disease (PD) was originally described by James Parkinson in 1817. PD is a progressive neurodegenerative disorder with an estimated prevalence of 0.3% in the U.S. population (Rao et al., 2006). The Rotterdam Study revealed a prevalence increasing to 4 to 5% in subjects older than 85 years (de Lau et al., 2004).

Pesticides (Dick, 2006) and increased body mass index are probably associated with a higher risk of PD (Hu et al., 2006). By contrast, coffee and tea drinking might be associated with a lower risk (Hu et al., 2007).

The majority of PD cases are sporadic. However, a classical Mendelian pattern of inheritance has been demonstrated in several families. Several genes linked to familial forms of disease (encoding alpha-synuclein, parkin, DJ-1, PINK-1, and LRRK2) have been identified. These families show either recessive or dominant inheritance patterns and may exhibit typical and/or atypical symptoms, with an age of onset extending from the second to the sixth decade (Thomas and Beal, 2007; MacInerney-Leo et al., 2005).

5.2.1   Motor Tests

Distal resting tremor of 3 to 6 Hz, rigidity (sustained increase of resistance throughout the range of passive movement about a joint), bradykinesia, and asymmetrical onset are cardinal signs. Other well-known signs include late-onset postural instability and micrographia. Response to an adequate therapeutic challenge of levodopa or a dopamine agonist is one of the key features for the diagnosis (guidelines for the diagnosis of Parkinson's disease, 2003). Parkinsonian patients often present an abnormal posture called camptocormia (ranging from mild to severe), which is characterized by an excessive anterior flexion of the spine (Bonneville et al., 2008). Bradykinesia is characterized by a difficulty to perform rapid and sequential tapping of the fingers or the heels, difficulty twiddling or circling the hands rapidly around each other in front of the body, reduced arm swing on the affected side during deambulation. Small, shuffling steps and troubles in initiating ambulation are common. Involuntary acceleration of gait results in the typical festinating gait. Heel-to-toe ambulation is impaired (Rao et al., 2006). Rigidity of the muscles on passive movement is characteristic of PD. Unlike spasticity (resulting from upper motor neuron lesions), the passive movement of the joints reveals continuous resistance throughout the entire range of motion, the so-called lead-pipe rigidity. In addition, patients with PD may show a cogwheel type of rigidity. Rigidity and cogwheel phenomenon may be triggered or enhanced by contralateral movement (Guttman et al., 2003).

Typically, the parkinsonian tremor is asymmetrical, at least initially, and affects the upper limb before involving the ipsilateral leg after a period of about 2 years. Tremor of the lips, jaw, or tongue may also occur. Head or voice tremor is rare, unlike in ET. A "pill-rolling" resting tremor is characteristic of PD. Postural tremor is also present in most cases. In many patients, two separate peaks of tremor may be distinguished by spectral analysis, and this pattern is considered very suggestive of basal ganglia disease by some authors (Habib-ur-Rehman, 2000). Kinetic tremor has been also described in PD, albeit uncommonly (Kraus et al., 2006). Patients affected by a Parkinson-plus syndrome, in particular, multiple system atrophy (MSA), often exhibit a slight to marked kinetic tremor during the course of the disease.

A study on the relation between center of pressure excursions and postural/resting limb tremor in PD patients, pointed out that the amplified tremor in PD can manifest itself in the center of pressure dynamics and influence postural stability (Kerr et al., 2008).

5.2.2 Nonmotor Tests
Decreased scores in cognitive tests are associated with greater impairment in motor performances (Verbaan et al., 2007). Among the symptoms of autonomic failure, orthostatic dizziness, bladder dysfunction, and constipation are considered to have great impact on daily life (Magerkurth et al., 2005). A decreased olfactory function has been reported.

5.2.3 Neuropathology
PD is due to the degeneration of dopaminergic neurons. Oxidative stress and mitochondrial dysfunction have been implicated as important contributors to neuronal death in substantia nigra (SN) of PD (Fukae et al., 2007). However, other neurotransmitter systems such as the serotonergic

coeruleus-subcoeruleus complex

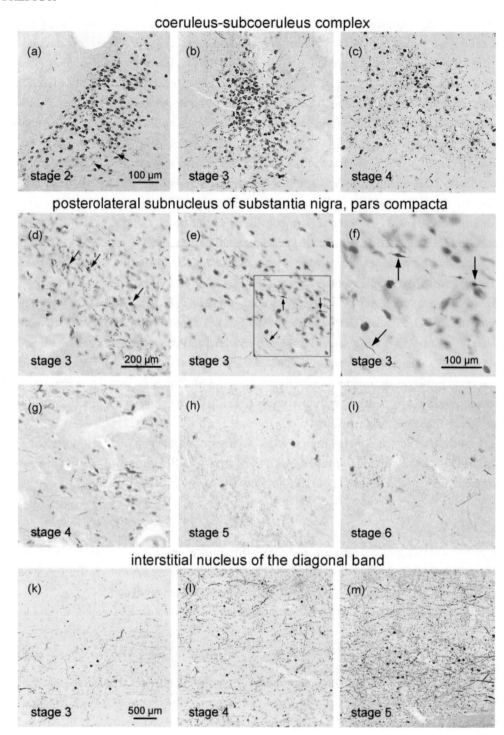

posterolateral subnucleus of substantia nigra, pars compacta

interstitial nucleus of the diagonal band

system also show signs of degeneration (Scholtissen et al., 2006). Progressive loss of SN neurons and presence of Lewy bodies are essential neuropathological features of PD. Recent neuropathological studies suggest that nigral degeneration is only part of a more extended brain degeneration that starts in the medulla oblongata and then spreads to the mesencephalon and cerebral cortex (Figure 5.3). Correspondingly, the clinical symptoms occurring in PD go far beyond parkinsonism. As mentioned earlier, autonomic dysfunction, olfactory disturbances, depression, and even dementia are frequently encountered in PD (Klockgether, 2004).

5.2.4 Differential Diagnosis

A list of extrapyramidal syndromes associated with tremor is given in Table 5.2. Parkinson plus syndromes, mainly progressive supranuclear palsy (PSP), multiple system atrophy (MSA), corticobasal ganglionic degeneration (CBD), Lewy body disease (LBD), vascular Parkinsonism, and Parkinsonism with no clear etiology are distinct from PD (Thanvi et al., 2005; Weiner, 2005). Parkinsonism on one side of the body in association with a contralateral atrophy of brain structures points towards the hemiparkinsonism–hemiatrophy syndrome (HPHA; Klawans, 1981). These patients are relatively young, exhibit a unilateral rest tremor, and may have a history of dystonia.

Data collected in an Italian center (over 2900 parkinsonian patients) showed that PD accounts for no more than 70% of all parkinsonisms (Pezzoli et al., 2004). Clinicopathological studies based on brain bank material from the UK and Canada have shown that clinicians diagnose the disease incorrectly in about 25% of patients (Tolosa et al., 2006).

FIGURE 5.3: Development of PD-related brain stem lesions. (a) The coeruleus–subcoeruleus complex is mildly involved at stage 2 [arrows point to Lewy neurites (LNs)]. (b) Severely affected at stage 3 (note the increase in LNs). (c) Severely depleted of melanin-containing projection neurons at stage 4. Bar in (a) is valid for (b and c). (d) The posterolateral and, thereafter, the posteromedial subnuclei of the substantia nigra, pars compacta, become first involved at stage 3 [arrows point to Lewy bodies (LBs) in melanized neurons]. Elongated LNs with spindle-shaped enlargements at irregular intervals (arrows) occur in the vulnerable subnuclei [the framed portion of (e) appears at higher magnification in (f)]. (g) Loss of melanized nerve cells is severe at stage 4. The vulnerable subnuclei are almost depleted of projection neurons at stages 5–6 (h and i). It is difficult to find surviving LB-containing nerve cells in these portions of the substantia nigra. Bar in (d) is valid for (e) and (g–i). (k) Like the substantia nigra, the interstitial nucleus of the diagonal band begins to develop LNs and LBs at stage 3. The density of both LNs and LBs increases in the following stages 4–5 (l–m). Bar in (k) is valid for (l and m). α-Synuclein immunoreactions, 100-μm thick PEG-sections. From Braak et al. (2003), with permission from Elsevier.

TABLE 5.2: Extrapyramidal syndromes associated with tremor
Parkinson's disease
"Parkinson-plus" (striatonigral degeneration SND, progressive supranuclear palsy PSP, multiple system atrophy MSA, olivopontocerebellar atrophy OPCA)
Neuroleptic-induced parkinsonism
Stroke in basal ganglia
Wilson's disease
Dystonia
Normopressure hydrocephalus
Midbrain lesions
Huntington's disease
Repeated head trauma (boxers, "punch-drunk syndrome")
Intoxication (CO, cyanide, manganese)
Spinocerebellar ataxias
Hallevorden–Spatz disease
Basal ganglia calcifications (hyperparathyroidism)
Alzheimer's disease
Parkinsonism dementia complex
Psychogenic

5.2.5   Neuroimaging

Positron emission tomography (PET) and single-photon emission computed tomography (SPECT) allow the in vivo assessment of the neurochemical, hemodynamic, or metabolic consequences of the degeneration of the nigrostriatal dopaminergic system in PD. The extent of striatal dopaminergic denervation can be quantified with radiotracers. Striatal uptake is markedly decreased in PD, affecting more the putamen than the caudate nucleus. This is inversely correlated with the severity of motor signs and with the duration of the disease (Thobois et al., 2001).

Hyperactivation in the cerebellum and motor cortex and hypoactivation in the basal ganglia in patients with PD have been described by neuroimaging studies. Recent evidence support the hypothesis that hyperactivation in the ipsilateral cerebellum is a compensatory mechanism for the defective basal ganglia. Moreover, hyperactivation in the contralateral primary motor cortex might not be just a compensatory response but seems directly related to upper limb rigidity (Yu et al., 2007).

5.3   CEREBELLAR DISEASES

Pathological conditions usually affect cerebellar function through several mechanisms: decrease in blood flow, edema, mechanical compression, invasion of cerebellar parenchyma, inflammatory and/or immune processes, and direct cytotoxic effect (Manto, 2002). The entities most commonly encountered in daily practice are cerebellar stoke, neurodegenerative ataxias (dominant spinocerebellar ataxias, Friedreich's ataxias, and other recessive forms), paraneoplastic cerebellar syndromes, cerebellitis, multiple sclerosis, multiple systemic atrophy (MSA), posttraumatic, metabolic, or toxic cerebellar ataxias, psychogenic ataxias.

Facing cerebellar disorders, some general principles should to be kept in mind:

1. focal cerebellar lesion located laterally generate signs ipsilaterally (see Chapter 2);
2. diffuse cerebellar disease, such as degenerative ataxias, usually cause relatively symmetric deficit;
3. cerebellar signs due to nonprogressive disease tend to undergo attenuation with time;
4. symptoms suggestive of cerebellar damage may be encountered in patients presenting lesions along the efferent or afferent cerebellar pathways outside the cerebellum (Manto, 2002).

As a rule, the symptoms of cerebellar disorders are influenced more by the location and the rate of progression of the disease than by the underlying pathology. As a consequence, an infection and a tumor, for instance, may produce similar cerebellar syndrome if they share the same location and rate of progression (Manto, 2002). The most common symptoms are gait difficulties, unsteadiness, headache, nausea/vomiting, dizziness, clumsiness in the limbs, speech difficulties, tremor, blurred vision, diplopia, feebleness, and sensory complaints.

Clinical signs may be a clue to the cerebellar area affected. Oculomotor deficits associated with ataxia of stance and gait indicate a *middle zone disease*. *Lateral lesions* are more likely to produce a combination of limb dysmetria (Figure 5.4), kinetic tremor, hypotonia, dysdiadochokinesia, impaired check, and excessive rebound. Isolated dysarthria has been reported in *intermediate zone* damage. Moreover, cerebellar dysarthria—characterized by scanning speech—is also observed in lesions located in the *cerebellar hemispheres* (Manto, 2002).

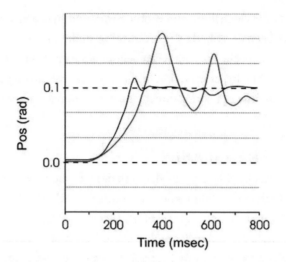

FIGURE 5.4: Overshoot in a cerebellar patient. The figure illustrates a typical cerebellar hypermetria (blue line) of the right hand in a right-handed patient presenting a severe cerebellar atrophy and exhibiting clinically a marked axial and appendicular cerebellar syndrome with oculomotor ataxia, dysmetria of limbs, and gait ataxia. The hypermetric movement is followed by attempts to reach the aimed target. Normometria in a control subject (black line). Aimed target: 0.1 rad.

Tremor associated with cerebellar diseases is mainly a low-frequency kinetic tremor. Many cerebellar patients exhibit a postural tremor. Isometric tremor as well as titubation may occur (Rondot and Bathien, 1995). The tremor may be bilateral or most often ipsilateral to the cerebellar lesion. In case of postural tremor, eye closure and body displacements enhance the oscillations. Atrophy of the anterior lobe of the cerebellum may be associated with a very suggestive 3-Hz leg tremor (Figure 5.5). The majority of these patients have a history of alcohol abuse (see also Section 4 below).

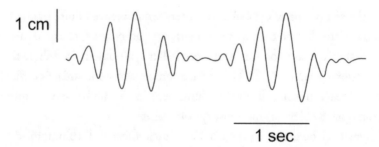

FIGURE 5.5: A 3-Hz leg tremor in a patient presenting atrophy of the anterior lobe of the cerebellum secondary to excessive alcohol consumption. Tremor is recorded during a sustained leg elevation with 90° flexion of the knee and hip joints, while the patient is supine. Tremor has a typical waxing and waning amplitude with a spindle-like aspect and predominates in one leg. Recordings from the right ankle.

TABLE 5.3: Classes of drugs and toxic agents inducing a postural and/or kinetic tremor				
A. Anticonvulsants: Phenytoin Carbamazepine overdose	C. Other drugs: Lithium salts Amiodarone Procainamide Cyclosporin and calcineurin inhibitors Bismuth Bromides/ bromvalerylurea Mefloquine Isoniazid Lindane Perhexiline maleate Metronidazole Glycoprotein IIb/IIIa inhibitors Nicotine	D. Drug abuse and addiction: Cocaine Heroin Phencyclidine Ceremonial herbs (kava) Methadone	F. Toluene/ Benzene Derivatives	J. Saxitoxin (shellfish poisoning)
			G. Carbon monoxide	K. Cyanide
			H. Chemical weapons: Diphenylarsinic acid poisoning	L. Complications following scorpion sting
B. Antineoplastics: 5-FU (5-fluorouracil) Ara-C cytosine (arabinoside) Methotrexate Capecitabine Epothilone D		E. Metals: Mercury Lead Manganese Aluminium Thallium Germanium Uranium Vanadium	I. Insecticides/ herbicides Chlodecone: Paraquat Phosphin Carbon disulfide	

5.3.1   Cerebellar Syndrome Induced by Drugs or Toxic Agents

As mentioned in Chapter 4, many drugs may enhance tremor. We summarize in Table 5.3 the classes of drugs and toxic agents which may induce cerebellar syndromes presenting with postural and/or kinetic tremor.

5.4   MISCELLANEOUS

5.4.1   Tremor of the Globe

Tremor of the eyes has frequencies varying from 2 to 70 Hz because of the mechanical properties of the eye.

5.4.2 Jaw Tremor

Jaw tremor can be seen in various neurological disorders such as ET, PD, dystonia, branchial my-oclonus, hereditary geniospasm, task-specific tremor, Whipple's disease (showing oculomasticatory myorhythmia: rhythmic contractions in the 1- to 3-Hz frequency range), as well as in normal situations such as shivering (hypothermia), and subclinical physiological jaw tremor (Gonzalez-Alegre et al., 2006). Frequency of jaw tremor is usually lower than 12 Hz.

5.4.3 Voice Tremor

Tremulousness of speech is not disease specific. It is encountered under physiological circumstances ("vibrato"; 4–7 Hz), in ET and PD (frequency 4–7 Hz), in amyotrophic lateral sclerosis (ALS). Speech analysis in ALS patients reveals tremor in about 60% of cases. Vocal cords display flaccid weakness, palatal/labial weakness. Tongue fasciculations may be observed because of denervation and loss of motor units, particularly in the bulbar form of ALS which affects predominantly cranial nerves (see Chapter 2). A rapid vocal flutter (7–10 Hz) may occur also (Aronson et al., 1992). Hypernasal resonation and imprecise articulation are common.

5.4.4 Multiple Sclerosis

Multiple sclerosis (MS) is a demyelinating disease of the central nervous system. Tremor is estimated to occur in about 25% to 60% of patients with MS (Koch et al., 2007). In MS, the most prevalent forms of tremor are postural tremor and kinetic tremor. Tremor of the head, neck, vocal cords, trunk, and limbs have all been described in association with MS. A true rest tremor is unusual (Koch et al., 2007). A case of Holmes tremor has been described recently (Yerdelen et al., 2008). So far, there are no reports of tremor of the palate, tongue, or jaw (Alusi et al., 1999).

5.4.5 Dystonic Tremor

Dystonic tremor occurs in generalized, segmental, and focal dystonias. In primary dystonia, tremor may anticipate the occurrence of clear signs of the disease. Secondary criteria like persistent focal tremor in one extremity, jerky and irregular tremors, gestes antagonistes, or selective responsiveness to antidystonic therapeutic agents are helpful to demarcate this entity. The occurrence of dystonic tremor in symptomatic cases is often observed after thalamic lesions, peripheral injuries, or in dystonia associated with sympathetic reflex dystrophy (Deuschl, 2003; see below).

Torticollis is the most common manifestation of dystonia (Jankovic et al., 1991). Even in the first description by Oppenheim, tremor was mentioned as a symptom associated with dystonic movements (Dubinsky, 1995). Isolated tremor of the trunk or head, especially of slow frequency (2–5 Hz; in the case of the head in a "no–no" direction), may be the initial manifestation of focal dystonia (Rivest and Marsden, 1990). A prospective study on 114 patients of cervical dystonia (CD) revealed that head tremor is common in CD and is more commonly associated with hand tremor (Pal et al., 2000).

The pathophysiology is not fully established (see Chapter 7). Impaired reciprocal inhibition and disturbances in other inhibitory reflex pathways at various levels are incriminated.

5.4.6   Orthostatic Tremor

Orthostatic tremor (OT) is a rare disorder of middle-aged or elderly people. OT is likely under-diagnosed. On average, there is a diagnostic delay of 5.7 years. In most cases, the condition remains relatively unchanged over the years, but symptoms may also spread proximally to involve the trunk and arms (Gerschlager et al., 2004). Cases of OT in brothers and in monozygotic twins were recently reported (Fischer et al., 2007; Contarino et al., 2006).

OT is characterized by unsteadiness on standing, which remits on walking or sitting (Piboolnurak et al., 2005). Lifting the standing patient off the ground abolishes the tremor. When walking, tremor disappears from the nonweight-bearing limb, but may persist in the weight-bearing leg and in truncal muscles (Heilman, 1984). Thereby, patients stand on a wide base but walk normally. Only a fine ripple of muscle activity is visible (Britton et al., 1992). Patients may describe a shaking sensation or cramps of the lower extremities upon standing. Typically, they are not able to stand in place for a long time because of unstability and the sensation of fatigue. A typical 14- to 18-Hz tremor of the lower limbs is pathognomonic (see Figure 4.6). Surface or needle EMG confirms the diagnosis.

Isometric contraction of arm and leg muscles triggers a high-frequency tremor in these patients (Boroojerdi et al., 1999). A single case report of orthostatic jaw tremor has been described (Schrag et al., 1999).

OT is divided into primary orthostatic tremor (POT) characterized by the absence of lesions detectable with imaging techniques and secondary orthostatic tremor (SOT), a symptomatic condition that has been described in hydrocephalus because of nontumoral aqueduct stenosis, in relapsing polyradiculoneuropathy (Gabellini et al., 1990), in pontine lesions (Benito-Leon et al., 1997), after head trauma (Sanitate and Meerschaert, 1993), as a paraneoplastic syndrome associated with small cell lung cancer (Gilhuis et al., 2005), in cerebellar diseases, and in the case of thiamine deficiency (Nasrallah and Mitsias, 2007). The possibility that hyperthyroidism might trigger OT has been raised recently (Tan et al., 2008).

5.4.7 Primary Writing Tremor

This is a task-specific tremor, mainly asymmetrical and focal, that interferes with handwriting. This tremor is often independent of posture or goal-directed motion. A bilateral presentation has been reported (Jimenez-Jimenez et al., 1998). Usually, it occurs alone. However, a concomitant postural tremor can be observed in some cases (Pahwa, 1995). Coexistence of tremor with dystonia has also been described (Elble et al., 1990). Although most of the patients have a negative family history for movements disorders, some relatives may be affected by ET, PD, (Pahwa in Hand book of tremor disorders), or dystonia (Hayashi and Koide, 1997).

Two subgroups are defined, depending on whether tremor appears during writing (type A: *task-induced tremor*) or according to the hand position used in writing (type B: *position-sensitive tremor*). The two groups differ also for the presence (type B) or absence (type A) of co-contracting EMG pattern and tremor induced by tendon taps to the volar aspect of the wrist (Bain et al., 1995).

Patients affected present significantly reduced writing speeds. A 4.1- to 7.3-Hz rhythmical activity predominantly in the intrinsic hand and forearm muscles is detected by polymyography performed during writing. A higher range of median frequency of 5.5 Hz has been detected by accelerometry. Normal subjects write with a median of 4.6 Hz (Bain et al., 1995). Rothwell and colleagues described a patient who complained of jerking of the right forearm on writing (Rothwell et al, 1979). In this patient, the active pronation of the arm produced several beats of pronation/supination tremor. Moreover, tendon taps to the volar surface of the wrist, to the finger extensors, and to the pectoralis major as well as forcible supination of the wrist (delivered by a torque motor) elicited bursts of tremor.

5.4.8 Complex Regional Pain Syndrome

This descriptive term refers to patients complaining of pain and presenting combinations of sensory, motor, and circulatory manifestations. The term algodystrophy is also related to this pathological condition. Complex regional pain syndrome (CRPS) has been classified in type I (formely "reflex sympathetic dystrophy" because of the presumed key role of the sympathetic nervous system) and type II (formerly "causalgia") occurring after partial injury of a peripheral nerve. The peripheral trauma is sometimes very minor, such as a simple foot sprain. CRPS is distinct from the syndrome of "painful legs and moving toes," for which a history of peripheral trauma or peripheral nerve dysfunction may occur (Spillane et al., 1971).

Tremor, spasms, dystonic deformities, and exaggerated tendon reflexes may develop in relation to CRPS. Postural, kinetic tremor, and enhanced postural tremor have been reported, usually with frequencies in the 3- to 6-Hz range (Verdugo and Ochoa, 2000). A prospective study on 145 patients affected by CRPS revealed motor disturbances in 97% of cases (Birklein et al., 2000).

5.4.9   Fragile-X-Associated Tremor/Ataxia Syndrome

Fragile-X-associated tremor/ataxia syndrome (FXTAS) is an adult-onset neurodegenerative disorder that affects carriers (principally males) of premutation alleles (55–200 CGG repeats) of the fragile X mental retardation 1 (FMR1) gene. Some adult male premutation carriers develop a neurological syndrome involving intention tremor, ataxia, dementia, parkinsonism, and autonomic dysfunction (Hagerman and Hagerman, 2004; Greco et al., 2006). Female carriers of the FMR1 premutation who presented with symptoms of tremor and ataxia have been described also. None of the women had dementia (Hagerman et al., 2004).

5.4.10   Familial Cortical Myoclonic Tremor with Epilepsy

The acronym FCMTE (familial cortical myoclonic tremor with epilepsy) has been proposed recently on the basis of both clinical and electrophysiological criteria: irregular postural myoclonic tremor of the distal limbs, familial history of epilepsy, autosomal dominant inheritance, and a rather benign outcome (Regragui et al., 2006). The diagnosis is confirmed by electrophysiological features suggesting a cortical reflex myoclonus (enhanced C-reflex, giant somatosensory-evoked potentials SEPs, premyoclonus cortical spikes detected by the back-averaging method), and response to antiepileptic drugs. The genetic analysis of these families shows a linkage to chromosome 8q24 for Japanese families and a linkage to chromosome 2p for Italian families (Regragui et al., 2006). The similarities of this syndrome with myoclonic epilepsies argue for an abnormality of a gene encoding ion channels.

5.4.11   Posttraumatic Tremor

As mentioned in Chapter 4, tremor may also rise as a consequence of central or peripheral trauma. Holmes tremor is an example of postural symptomatic tremor because of central trauma (Bhidayasiri, 2005), including brainstem hemorrhage (Raina et al., 2007). A 46-year-old man who developed unilateral dystonic tremor 13 years after contralateral combat-induced head trauma has been reported (Newmark and Richards, 1999). Hashimoto et al. (2002) have described a 55-year-old man who developed, 7 days after a neck and spine trauma, a disabling coarse postural tremor in the right upper extremity, along with numbness in the ulnar side of the right forearm. A C7–C8 radiculopathy was detected, while brain and cervical spine MRI were both unremarkable. Loading and ischemic nerve block reduced the tremor frequency, in agreement with a mechanical reflex mechanism underlying the tremor (see Chapter 1). Costa and coworkers have described a patient with proximal right upper limb tremor secondary to direct peripheral nerve lesion caused by prior thoracic surgery (Costa et al., 2006).

Magnetic resonance evidence of lesions in the dentatoe–thalamic pathway was described in patients with severe posttraumatic tremor (Krauss et al., 1995). Lesions may involve the Guillain–Mollaret triangle (see Chapter 2).

A contribution of predisposing factors to the pathogenesis of the posttraumatic tremor has been proposed (Jankovic and Van der Linden, 1988).

5.4.12 Tremor in Postinfectious Syndromes

West Nile virus is considered the most common arboviral infection in North America. Meningoencephalitis, acute flaccid paralysis, and movement disorders including tremor, myoclonus, or parkinsonism are the main clinical presentations. Maculopapular rash and tremor were demonstrated as independently associated with the infection (Tilley et al., 2007). Beside the multiple somatic complaints and abnormalities in motor skills and executive functions, tremor is reported as a common long-term problem among patients who have had West Nile virus infection (Carson et al., 2006).

Postinfectious encephalopathy with tremor may develop depending on possible lesions in the brain in cases of cytomegalovirus infection, toxoplasmosis, HIV infection, and malaria (Coelho et al., 1996; Mizuno et al., 2006; Cardoso, 2002). A midbrain abscess may generate Holmes tremor (Pezzini et al., 2002).

5.4.13 Drug-Induced Tremor

Tremor is the most frequent drug-induced movement disorder. Drug intake (and true or 'hidden' neuroleptic or psychoactive drug intake, see Tables 4.2 and 5.3) should be suspected as a possible cause of movement disorders. Drug-induced acute movement disorders usually disappear spontaneously after withdrawal of the suspected drug (Montastruc and Durrieu, 2004). The delay for disappearance of tremor may vary from days to years.

The most common drug-induced tremor is enhanced physiological tremor after use of sympathomimetics such as pseudoephedrine, bronchodilators, or theophylline, and antidepressants such as tricyclic agents or fluoxetine. Approximately 25% of patients taking valproic acid therapy (taken mainly for treatment of seizures or headaches) exhibit a postural tremor 3 to 12 months after starting therapy. Lowering the dosage decreases the intensity of tremor. Lithium induces a fine postural tremor of the hands (frequency 8–12 Hz). Lithium toxicity may cause permanent damage to the cerebellum that precipitates postural and intention tremors. Amiodarone may cause a dose-dependent reversible neurologic syndrome consisting of postural tremor, ataxia, and peripheral neuropathy. Symptoms develop in the first weeks of treatment and improve slowly after dose reduction or discontinuation. Neuroleptic agents such as haloperidol or dopamine-receptor-blocking drugs like metoclopramide may induce a parkinsonian tremor (Smaga, 2003). Anticalcic agents may also trigger a parkinsonian syndrome in the elderly.

5.4.14 Alcohol Abuse and Alcohol Withdrawal Syndrome

Alcohol is a major toxic agent for the brain, in particular for the cerebellum (Pentney, 2002). Chronic alcohol consumption leads to several neurological complications: cerebellar dysfunction (Gilman et al., 1981), painful small-fiber peripheral neuropathy (whose severity increases during alcohol withdrawal; Dina et al., 2008), impairment of cerebral vasoreactivity, thus eventually resulting in an increase in ischemic brain damage (Sun et al., 2008). During acute alcohol intoxication, patients exhibit slurred speech, gaze-evoked nystagmus, anterior–posterior oscillations in Romberg test, a broad-based ataxic gait. A 3-Hz postural leg tremor and kinetic tremor are frequently found in chronic alcoholic patients (Manto and Jacquy, 2002). One of the most severe consequences of alcoholism is the Wernicke–Korsakoff encephalopathy, historically described with confusion, ataxia, ophtalmoplegia, confabulation, and amnesic disturbances (Pearce, 2008).

A suddenly decrease in alcohol consumption or a sudden abstinence may lead to alcohol withdrawal syndrome's (AWS) comprising mild to moderate postural tremor, irritability, anxiety, or agitation. The most severe manifestations of withdrawal is characterized by delirium tremens, hallucinations, and seizures (Saitz, 1998).

5.4.15 Peripheral Neuropathy

Acute or chronic demyelinating (Guillain–Barre´ syndrome and Charcot–Marie–Tooth syndrome, respectively) and dysgammaglobulinemic neuropathies are the most common causes of tremor secondary to neuropathy. An action tremor resembling ET evolves. Postural tremor has been described in other disorders causing neuropathies including X-linked bulbospinomuscular atrophy, multifocal motor neuropathy, and human T lymphotropic virus 1-associated neuropathy (Habib-ur-Rehman, 2000). Tremor has been reported also in a few cases of peripheral neuropathy associated with diabetes, uremia, alcohol intake, and during treatment with drugs causing a peripheral neuropathy (Smith, 1995).

5.4.16 Psychogenic Tremor

Movement disorders—especially tremor, dystonia, and myoclonus—may be the presentation of a psychogenic disorder. Approximately 3% of patients followed in specialized clinics have a psychogenic disease which represents a conversion disorder or part of somatoform disorder (Reich, 2006). Tremor may affect any region of the body. Cases of psychogenic palatal tremor have been described

(Pirio Richardson et al., 2006). Comorbidity with other movement disorders is not exceptional. A sudden onset and the variability in tremor features are common (see also Chapter 4). The "coactivation sign of psychogenic tremor" and the absence of finger tremor suggest a psychogenic origin. Quantitative analysis of tremor shows decreasing amplitudes in most organic tremors when the extremity is loaded with additional weights. In contrast, tremor amplitude increases with added inertia for most of the cases with psychogenic tremor (Deuschl et al., 1998).

According to a modified Fahn's criteria, the diagnosis of psychogenic tremor is considered in the following conditions:

- exclusion of the major causes of symptomatic tremor
- exclusion of ET and PD
- absence of evidence for any other neurological disorders
- a period without tremor of at least 2 weeks during the observation period (Bhidayasiri, 2005)

5.5   ROLE OF BRAIN IMAGING IN THE DIFFERENTIAL DIAGNOSIS OF TREMOR

5.5.1   Computerized Tomography and Magnetic Resonance Imaging

Computerized tomography (CT scan) and magnetic resonance imaging (MRI) are useful for the diagnosis of tremors associated with to neurodegenerative diseases, vascular diseases, or posttraumatic tremor. *Brain MRI* has become a key tool for the morphological assessment of brain structures, being more sensitive and specific than the CT scan. For instance, brain MRI may detect an enlargement of the inferior olivary nucleus in palatal tremor as a result of a transsynaptic degeneration of the Guillain–Mollaret triangle (Yokota et al., 1989; Barron et al., 1982).

5.5.2   Positron Emission Tomography

Positron emission tomography (PET; see Chapter 7) with ^{18}F-Dopa and specific tracers to measure dopamine transporter (DAT; SPECT) binding provide an accurate measure of degeneration in the dopamine system in PD (Antonini et al., 2001). SPECT ligands provide a marker for presynaptic neuronal degeneration (Marshall et al., 2003). Low striatal uptake of the radioligand differentiates PD from ET (Benamer et al., 2000) and can be used to monitor disease severity (Breit et al., 2006). SPECT seems accurate in differentiating idiopathic PD (PD) from vascular parkinsonism and drug-induced parkinsonism (Vlaar et al., 2008). Moreover, PET might allow the selection of at-risk PD subjects (in early or preclinical stage) for neuroprotective intervention trials as well as for the evaluation of the effects of neurorestorative interventions in patients who are at an advanced stage (Bohnen and Frey, 2003).

5.5.3 Transcranial Sonography

Transcranial sonography assesses echogenicity of the substantia nigra as a characteristic of PD. This procedure might help to differentiate idiopathic PD from atypical parkinsonian syndromes at early stages (Gaenslen et al., 2008). Transcranial sonography is commonly used for stroke assessments.

REFERENCES

Alusi SH, Glickman S, Aziz TZ, Bain PG. Tremor in multiple sclerosis. *J Neurol Neurosurg Psychiatry.* 1999;66(2):131–134.

Antonini A, Moresco RM, Gobbo C, De Notaris R, Panzacchi A, Barone P, Calzetti S, Negrotti A, Pezzoli G, Fazio F. The status of dopamine nerve terminals in Parkinson's disease and essential tremor: a PET study with the tracer [11-C]FE-CIT. *Neurol Sci.* 2001;22(1):47–48. doi:10.1007/s100720170040

Aronson AE, Ramig LO, Winholtz WS, Silber SR. Rapid voice tremor, or "flutter," in amyotrophic lateral sclerosis. *Ann Otol Rhinol Laryngol.* 1992;101:511–518.

Bain PG, Findley LJ, Thompson PD, Gresty MA, Rothwell JC, Harding AE, Marsden CD. A study of hereditary essential tremor. *Brain.* 1994;117(Pt 4):805–824. doi:10.1093/brain/117.4.805

Bain PG, Findley LJ, Britton TC, Rothwell JC, Gresty MA, Thompson PD, Marsden CD. Primary writing tremor. *Brain.* 1995;118(Pt 6):1461–1472. doi:10.1093/brain/118.6.1461

Barron KD, Dentinger MP, Koeppen AH. Fine structure of neurons of the hypertrophied human inferior olive. *J Neuropathol Exp Neurol.* 1982;41:186–203.

Benamer TS, Patterson J, Grosset DG, Booij J, de Bruin K, van Royen E, Speelman JD, Horstink MH, Sips HJ, Dierckx RA, Versijpt J, Decoo D, Van Der Linden C, Hadley DM, Doder M, Lees AJ, Costa DC, Gacinovic S, Oertel WH, Pogarell O, Hoeffken H, Joseph K, Tatsch K, Schwarz J, Ries V. Accurate differentiation of parkinsonism and essential tremor using visual assessment of [123I]-FP-CIT SPECT imaging: the [123I]-FP-CIT study group. *Mov Disord.* 2000;15(3):503–510. doi:10.1002/1531-8257(200005)15:3<503::AID-MDS1013>3.0.CO;2-V

Benito-León J, Rodríguez J, Ortí-Pareja M, Ayuso-Peralta L, Jiménez-Jiménez FJ, Molina JA. Symptomatic orthostatic tremor in pontine lesions. *Neurology.* 1997;49(5):1439–1441.

Benito-León J, Louis ED. Essential tremor: emerging views of a common disorder. *Nat Clin Pract Neurol.* 2006;2(12):666–678. doi:10.1038/ncpneuro0347

Bhidayasiri R. Differential diagnosis of common tremor syndromes. *Postgrad Med J.* 2005;81:756–762. doi:10.1136/pgmj.2005.032979

Birklein F, Riedl B, Sieweke N, Weber M, Neundörfer B. Neurological findings in complex regional pain syndromes—analysis of 145 cases. *Acta Neurol Scand.* 2000;101(4):262–269. doi:10.1034/j.1600-0404.2000.101004262x./

Bohnen NI, Frey KA. The role of positron emission tomography imaging in movement disorders. *Neuroimaging Clin N Am.* 2003;13(4):791–803. doi:10.1016/S1052-5149(03)00096-0

Bonneville F, Bloch F, Kurys E, du Montcel ST, Welter ML, Bonnet AM, Agid Y, Dormont D, Houeto JL. Camptocormia and Parkinson's disease: MR imaging. *Eur Radiol.* 2008;18:1710–1719. doi:10.1007/s00330-008-0927-8

Boroojerdi B, Ferbert A, Foltys H, Kosinski CM, Noth J, Schwarz M. Evidence for a nonorthostatic origin of orthostatic tremor. *J Neurol Neurosurg Psychiatry.* 1999;66(3):284–288.

Braak H, Del Tredici K, Rüb U, de Vos RA, Jansen Steur EN, Braak E. Staging of brain pathology related to sporadic Parkinson's disease. *Neurobiol Aging.* 2003;24(2):197–211.

Breit S, Reimold M, Reischl G, Klockgether T, Wüllner U. [(11)C]D-Threo-methylphenidate PET in patients with Parkinson's disease and essential tremor. *J Neural Transm.* 2006;113(2):187–193.

Brennan KC, Jurewicz EC, Ford B, Pullman SL, Louis ED. Is essential tremor predominantly a kinetic or a postural tremor? A clinical and electrophysiological study. *Mov Disord.* 2002;17(2):313–316. doi:10.1002/mds.10003

Britton TC, Thompson PD, van der Kamp W, Rothwell JC, Day BL, Findley LJ, Marsden CD. Primary orthostatic tremor: further observations in six cases. *J Neurol.* 1992;239(4):20–217. doi:10.1007/BF00839142

Cardoso F. HIV-related movement disorders: epidemiology, pathogenesis and management. *CNS Drugs.* 2002;16(10):663–668. doi:10.2165/00023210-200216100-00002

Carson PJ, Konewko P, Wold KS, Mariani P, Goli S, Bergloff P, Crosby RD. Long-term clinical and neuropsychological outcomes of West Nile virus infection. *Clin Infect Dis.* 2006;43(6):723–730. doi:10.1086/506939

Coelho JC, Wiederkehr JC, Cat R, Carrero JE, de Oliveira ED, Campos AC, Cat I. Extrapyramidal disorder secondary to cytomegalovirus infection and toxoplasmosis after liver transplantation. *Eur J Pediatr Surg.* 1996;6(2):110–111.

Cohen O, Pullman S, Jurewicz E, Watner D, Louis ED. Rest tremor in patients with essential tremor: prevalence, clinical correlates, and electrophysiologic characteristics. *Arch Neurol.* 2003;60(3):405–410. doi:10.1001/archneur.60.3.405

Contarino MF, Welter ML, Agid Y, Hartmann A. Orthostatic tremor in monozygotic twins. *Neurology.* 2006;66(10):1600–1601. doi:10.1212/01.wnl.0000216263.23642.db

Costa J, Henriques R, Barroso C, Ferreira J, Atalaia A, de Carvalho M. Upper limb tremor induced by peripheral nerve injury. *Neurology.* 2006;67(10):1884–1886. doi:10.1212/01. wnl.0000244437.31413.2b

de Lau LM, Giesbergen PC, de Rijk MC, Hofman A, Koudstaal PJ, Breteler MM. Incidence of parkinsonism and Parkinson disease in a general population: the Rotterdam Study. *Neurology.* 2004;63(7):1240–1244.

Deng H, Le W, Jankovic J. Genetics of essential tremor. *Brain.* 2007;130(Pt 6):1456–1464.

Deuschl G, Bain P, Brin M. Consensus statement of the Movement Disorder Society on Tremor. Ad Hoc Scientific Committee. *Mov Disord.* 1998;13(3):2–23.

Deuschl G, Köster B, Lücking CH, Scheidt C. Diagnostic and pathophysiological aspects of psychogenic tremors. *Mov Disord.* 1998;13(2):294–302. doi:10.1002/mds.870130216

Deuschl G. Dystonic tremor. *Rev Neurol (Paris).* 2003;159(10, Pt 1):900–905.

Dick FD. Parkinson's disease and pesticide exposures. *Br Med Bull.* 2006;79–80:219–31. doi:10.1093/bmb/ldl018

Dina OA, Khasar SG, Alessandri-Haber N, Green PG, Messing RO, Levine JD. Alcohol-induced stress in painful alcoholic neuropathy. *Eur J Neurosci.* 2008;27(1):83–92.

Dubinsky RM. Tremor and dystonia. In: *Handbook of Tremor Disorders.* Findley LJ, Koller WC (eds). New York: Marcel Dekker, 1995.

Elble RJ, Higgins C, Leffler K, Hughes L. Factors influencing the amplitude and frequency of essential tremor. *Mov Disord.* 1994;9(6):589–596. doi:10.1002/mds.870090602

Elble RJ, Moody C, Higgins C. Primary writing tremor. A form of focal dystonia? *Mov Disord.* 1990;5(2):118–126. doi:10.1002/mds.870050205

Findley LJ, Koller WC. *Handbook of Tremor Disorders.* New York: Marcel Dekker, 1995.

Fischer M, Kreß W, Reiners K, Rieckmann P. Orthostatic tremor in three brothers. *J Neurol.* 2007;254(12):1759–1760. doi:10.1007/s00415-007-0647-z

Fukae J, Mizuno Y, Hattori N. Mitochondrial dysfunction in Parkinson's disease. *Mitochondrion.* 2007;7(1–2):58–62. doi:10.1016/j.mito.2006.12.002

Gabellini AS, Martinelli P, Gullì MR, Ambrosetto G, Ciucci G, Lugaresi E. Orthostatic tremor: essential and symptomatic cases. *Acta Neurol Scand.* 1990;81(2):113–117.

Gaenslen A, Unmuth B, Godau J, Liepelt I, Di Santo A, Schweitzer KJ, Gasser T, Machulla HJ, Reimold M, Marek K, Berg D. The specificity and sensitivity of transcranial ultrasound in the differential diagnosis of Parkinson's disease: a prospective blinded study. *Lancet Neurol.* 2008;7(5):417–424.

Gerschlager W, Münchau A, Katzenschlager R, Brown P, Rothwell JC, Quinn N, Lees AJ, Bhatia KP. Natural history and syndromic associations of orthostatic tremor: a review of 41 patients. *Mov Disord.* 2004;19(7):788–795. doi:10.1002/mds.20132

Gilhuis HJ, van Ommen HJ, Pannekoek BJ, Sillevis Smitt PA. Paraneoplastic orthostatic tremor associated with small cell lung cancer. *Eur Neurol.* 2005;54(4):225–226. doi:10.1159/000090715

Gilman S, Bloedel JR, Lechtenberg R. Disorders of the cerebellum. In: *Contemporary Neurology Series.* Davis FA (Ed). Philadelphia. 1981.

Gonzalez-Alegre P, Kelkar P, Rodnitzky RL. Isolated high-frequency jaw tremor relieved by botulinum toxin injections. *Mov Disord.* 2006;21(7):1049–1050. doi:10.1002/mds.20878

Greco CM, Berman RF, Martin RM, Tassone F, Schwartz PH, Chang A, Trapp BD, Iwahashi C, Brunberg J, Grigsby J, Hessl D, Becker EJ, Papazian J, Leehey MA, Hagerman RJ, Hagerman PJ. Neuropathology of fragile X-associated tremor/ataxia syndrome (FXTAS). *Brain.* 2006;129(Pt 1):243–255.

Guttman M, Kish SJ, Furukawa Y. Current concepts in the diagnosis and management of Parkinson's disease. *CMAJ.* 2003;168(3):293–301.

Habib-ur-Rehman. Diagnosis and management of tremor. *Arch Intern Med.* 2000;160(16):2438–2444. doi:10.1001/archinte.160.16.2438

Hagerman PJ, Hagerman RJ. Fragile X-associated tremor/ataxia syndrome (FXTAS). *Ment Retard Dev Disabil Res Rev.* 2004;10(1):25–30.

Hagerman RJ, Leavitt BR, Farzin F, Jacquemont S, Greco CM, Brunberg JA, Tassone F, Hessl D, Harris SW, Zhang L, Jardini T, Gane LW, Ferranti J, Ruiz L, Leehey MA, Grigsby J, Hagerman PJ. Fragile-X-associated tremor/ataxia syndrome (FXTAS) in females with the FMR1 premutation. *Am J Hum Genet.* 2004;74(5):1051–1056.

Hashimoto T, Sato H, Shindo M, Hayashi R, Ikeda S. Peripheral mechanisms in tremor after traumatic neck injury. *J Neurol Neurosurg Psychiatry.* 2002;73:585–587. doi:10.1136/jnnp.73.5.585

Hayashi M, Koide H. Idiopathic torsion dystonia and writing tremor within a family. *Brain Dev.* 1997;19(8):556–558. doi:10.1016/S0387-7604(97)00070-3

Heilman KM. Orthostatic tremor. *Arch Neurol.* 1984;41(8):880–881.

Hu G, Jousilahti P, Nissinen A, Antikainen R, Kivipelto M, Tuomilehto J. Body mass index and the risk of Parkinson disease. *Neurology.* 2006;67(11):1955–1959. doi:10.1212/01.wnl.0000247052.18422.e5

Hu G, Bidel S, Jousilahti P, Antikainen R, Tuomilehto J. Coffee and tea consumption and the risk of Parkinson's disease. *Mov Disord.* 2007;22(15):2242–2248. doi:10.1002/mds.21706

Italian Neurological Society; Italian Society of Clin Neurophysiol; Guidelines for the Treatment of Parkinson's Disease 2002. The diagnosis of Parkinson's disease. *Neurol Sci.* 2003;24(3):S157–S164.

Jankovic J, Van der Linden C. Dystonia and tremor induced by peripheral trauma: predisposing factors. *J Neurol Neurosurg Psychiatry*. 1988;51(12):1512–1519.

Jankovic J, Leder S, Warner D, Schwartz K. Cervical dystonia: clinical findings and associated movement disorders. *Neurology*. 1991;41(7):1088–1091.

Jimenez-Jimenez FJ, Cabrera-Valdivia F, Orti-Pareja M, Gasalla T, Tallon-Barranco A, Zurdo M. Bilateral primary writing tremor. *Eur J Neurol*. 1998;5(6):613–614. doi:10.1046/j.1468-1331.1998.560613.x

Jiménez-Jiménez FJ, de Toledo-Heras M, Alonso-Navarro H, Ayuso-Peralta L, Arévalo-Serrano J, Ballesteros-Barranco A, Puertas I, Jabbour-Wadih T, Barcenilla B. Environmental risk factors for essential tremor. *Eur Neurol*. 2007;58(2):106–113.

Kerr G, Morrison S, Silburn P. Coupling between limb tremor and postural sway in Parkinson's disease. *Mov Disord*. 2008;23:386–394. doi:10.1002/mds.21851

Klawans HL. Hemiparkinsonism as a late complication of hemiatrophy: a new syndrome. *Neurology*. 1981;31:625–628.

Klockgether T. Parkinson's disease: clinical aspects. *Cell Tissue Res*. 2004;318(1):115–120. doi:10.1007/s00441-004-0975-6

Koch M, Mostert J, Heersema D, De Keyser J. Tremor in multiple sclerosis. *J Neurol*. 2007;254(2): 133–145. doi:10.1007/s00415-006-0296-7

Kraus PH, Lemke MR, Reichmann H. Kinetic tremor in Parkinson's disease—an underrated symptom. *J Neural Transm*. 2006;113(7):845–853. doi:10.1007/s00702-005-0354-9

Krauss JK, Wakhloo AK, Nobbe F, Tränkle R, Mundinger F, Seeger W. Lesion of dentatothalamic pathways in severe post-traumatic tremor. *Neurol Res*. 1995;17(6):409–416.

Lou JS, Jankovic J. Essential tremor: clinical correlates in 350 patients. *Neurology*. 1991;41(2, Pt 1): 234–238

Louis ED, Zheng W, Jurewicz EC, Watner D, Chen J, Factor-Litvak P, Parides M. Elevation of blood beta-carboline alkaloids in essential tremor. *Neurology*. 2002;59(12):1940–1944.

Louis ED, Ottman R. Study of possible factors associated with age of onset in essential tremor. *Mov Disord*. 2006;21(11):1980–1986. doi:10.1002/mds.21102

McInerney-Leo A, Hadley DW, Gwinn-Hardy K, Hardy J. Genetic testing in Parkinson's disease. *Mov Disord*. 2005;20(1):1–10.

Magerkurth C, Schnitzer R, Braune S. Symptoms of autonomic failure in Parkinson's disease: prevalence and impact on daily life. *Clin Auton Res*. 2005;15(2):76–82. doi:10.1007/s10286-005-0253-z

Manto M. Clinical signs of cerebellar disorders. In: *The Cerebellum and Its Disorders*. Manto MU, Pandolfo M (eds). Cambridge, UK: Cambridge University Press, 2002.

Manto M, Jacquy J. Alcohol toxicity in the cerebellum. In: *Clinical Aspects.* Manto M.U. and Pandolfo M (eds). Cambridge, UK: Cambridge University Press, 2002, pp. 336–341.

Marshall V, Grosset D. Role of dopamine transporter imaging in routine clinical practice. *Mov Disord.* 2003;18(12):1415–1423. doi:10.1002/mds.10592

Mizuno Y, Kato Y, Kanagawa S, Kudo K, Hashimoto M, Kunimoto M, Kano S. A case of postmalaria neurological syndrome in Japan. *J Infect Chemother.* 2006;12(6):399–401.

Montastruc JL, Durrieu G. Drug-induced tremor and acute movement disorders. *Therapie.* 2004;59(1):97–103.

Nasrallah KM, Mitsias PD. Orthostatic tremor due to thiamine deficiency. *Mov Disord.* 2007;22(3):440–441. doi:10.1002/mds.21193

Newmark J, Richards TL. Delayed unilateral post-traumatic tremor: localization studies using single-proton computed tomographic and magnetic resonance spectroscopy techniques. *Mil Med.* 1999;164(1):59–64.

Pahwa R. Primary writing tremor. In: *Handbook of Tremor Disorders.* Findley LJ, Koller WC (eds). New York: Marcel Dekker, 1995.

Pal PK, Samii A, Schulzer M, Mak E, Tsui JK. Head tremor in cervical dystonia. *Can J Neurol Sci.* 2000;27(2):137–142.

Paulson GW. Benign essential tremor: features that aid in diagnosis. *Postgrad Med J.* 1982;71(1):105–107.

Pearce JM. Wernicke–Korsakoff encephalopathy. *Eur Neurol.* 2008;59(1–2):101–104. doi:10.1159/000109580

Pentney R. Alcohol toxicity in the cerebellum: fundamental aspects. In: *The Cerebellum and Its Disorders.* Manto MU and Pandolfo M (eds). Cambridge, UK: Cambridge University Press, 2002, pp. 327–339.

Pezzini A, Zavarise P, Palvarini L, Viale P, Oladeji O, Padovani A. Holmes' tremor following midbrain Toxoplasma abscess: clinical features and treatment of a case. *Parkinsonism Relat Disord.* 2002;8(3):177–180. doi:10.1016/S1353-8020(01)00013-X

Pezzoli G, Canesi M, Galli C. An overview of parkinsonian syndromes: data from the literature and from an Italian data-base. *Sleep Med.* 2004;5(2):181–187. doi:10.1016/j.sleep.2003.10.009

Piboolnurak P, Yu QP, Pullman SL. Clinical and neurophysiologic spectrum of orthostatic tremor: case series of 26 subjects. *Mov Disord.* 2005;20(11):1455–1461. doi:10.1002/mds.20588

Pirio Richardson S, Mari Z, Matsuhashi M, Hallett M. Psychogenic palatal tremor. *Mov Disord.* 2006;21(2):274–276. doi:10.1002/mds.20731

Raina GB, Velez M, Pardal MF, Micheli F. Holmes tremor secondary to brainstem hemorrhage responsive to levodopa: report of 2 cases. *Clin Neuropharmacol.* 2007;30(2):95–100. doi:10.1097/01.wnf.0000240957.56939.e6

Regragui W, Gerdelat-Mas A, Simonetta-Moreau M. Cortical tremor (FCMTE: familial cortical myoclonic tremor with epilepsy). *Neurophysiol Clin.* 2006;36(5–6):345–349. doi:10.1016/j.neucli.2006.12.005

Rao SS, Hofmann LA, Shakil A. Parkinson's disease: diagnosis and treatment. *Am Fam Physician.* 2006;74(12):2046–2054.

Reich SG. Psychogenic movement disorders. *Semin Neurol.* 2006;26(3):289–296. doi:10.1055/s-2006-947276

Rivest J, Marsden CD. Trunk and head tremor as isolated manifestations of dystonia. *Mov Disord.* 1990;5(1):60–65. doi:10.1002/mds.870050115

Rondot P, Bathien N. Cerebellar tremors: physiological basis and treatment. In: *Handbook of Tremor Disorders.* Findley LJ, Koller WC (eds). New York: Marcel Dekker, 1995.

Rothwell JC, Traub MM, Marsden CD. Primary writing tremor. *J Neurol Neurosurg Psychiatry.* 1979;42(12):1106–1114.

Saitz R. Introduction to alcohol withdrawal. *Alcohol Health Res World.* 1998;22(1):5–12.

Sanitate SS, Meerschaert JR. Orthostatic tremor: delayed onset following head trauma. *Arch Phys Med Rehabil.* 1993;74(8):886–889.

Scholtissen B, Verhey FR, Steinbusch HW, Leentjens AF. Serotonergic mechanisms in Parkinson's disease: opposing results from preclinical and clinical data. *J Neural Transm.* 2006;113(1):59–73.

Schrag A, Bhatia K, Brown P, Marsden CD. An unusual jaw tremor with characteristics of primary orthostatic tremor. *Mov Disord.* 1999;14(3):528–530. doi:10.1002/1531-8257(199905)14:3<528::AID-MDS1029>3.0.CO;2-E

Smaga S. Tremor. *Am Fam Physician.* 2003;68(8):1545–1552.

Smith IS. Tremor in peripheral neuropathy. In: *Handbook of Tremor Disorders.* Findley LJ, Koller WC (eds). New York: Marcel Dekker, 1995.

Spillane JD, Nathan PW, Kelly RE, Marsden CD. Painful legs and moving toes. *Brain.* 1971;94:541–556. doi:10.1093/brain/94.3.541

Stockner H, Sojer M, K KS, Mueller J, Wenning GK, Schmidauer C, Poewe W. Midbrain sonography in patients with essential tremor. *Mov Disord.* 2007;22(3):414–417. doi:10.1002/mds.21344

Sun H, Zhao H, Sharpe GM, Arrick DM, Mayhan WG. Effect of chronic alcohol consumption on brain damage following transient focal ischemia. *Brain Res.* 2008;1194:73–80.

Sur H, Ilhan S, Erdo an H, Oztürk E, Ta demir M, Börü UT. Prevalence of essential tremor: a door-to-door survey in Sile, Istanbul, Turkey. *Parkinsonism Relat Disord.* 2008. doi:10.1016/j.parkreldis.2008.03.009

Tan EK, Lo YL, Chan LL. Graves disease and isolated orthostatic tremor. *Neurology.* 2008;70(16, Pt 2):1497–1498. doi:10.1212/01.wnl.0000310405.36026.92

Thanvi B, Lo N, Robinson T. Vascular parkinsonism—an important cause of parkinsonism in older people. *Age Ageing.* 2005;34(2):114–119. doi:10.1093/ageing/afi025

Thanvi B, Lo N, Robinson T. Essential tremor-the most common movement disorder in older people. *Age Ageing.* 2006;35(4):344–349. doi:10.1093/ageing/afj072

Thobois S, Guillouet S, Broussolle E. Contributions of PET and SPECT to the understanding of the pathophysiology of Parkinson's disease. *Neurophysiol Clin.* 2001;31(5):321–340.

Thomas B, Beal MF. Parkinson's disease. *Hum Mol Genet.* 2007;16(Spec No. 2):R183–R194.

Tilley PA, Fox JD, Jayaraman GC, Preiksaitis JK. Maculopapular rash and tremor are associated with West Nile fever and neurological syndromes. *J Neurol Neurosurg Psychiatry.* 2007; 78(5):529–531.

Tolosa E, Wenning G, Poewe W. The diagnosis of Parkinson's disease. *Lancet Neurol.* 2006;5(1):75–86. doi:10.1016/S1474-4422(05)70285-4

Verbaan D, Marinus J, Visser M, van Rooden SM, Stiggelbout AM, Middelkoop HA, van Hilten JJ. Cognitive impairment in Parkinson's disease. *J Neurol Neurosurg Psychiatry.* 2007;78(11):1182–1187. doi:10.1136/jnnp.2006.112367

Verdugo RJ, Ochoa JL. Abnormal movements in complex regional pain syndrome: assessment of their nature. *Muscle Nerve.* 2000;23(2):198–205.

Vlaar AM, de Nijs T, Kessels AG, Vreeling FW, Winogrodzka A, Mess WH, Tromp SC, van Kroonenburgh MJ, Weber WE. Diagnostic value of ^{123}I-ioflupane and ^{123}I-iodobenzamide SPECT scans in 248 patients with parkinsonian syndromes. *Eur Neurol.* 2008;59(5): 258–266. doi:10.1159/000115640

Weiner WJ. A differential diagnosis of Parkinsonism. *Rev Neurol Dis.* 2005;2(3):124–131.

Yerdelen D, Karatas M, Goksel B, Yildirim T. A patient with multiple sclerosis presenting with Holmes' tremor. *Eur J Neurol.* 2008;15(1):e2–e3.

Yokota T, Hirashima F, Furukawa T, Tsukagoshi H, Yoshikawa H. MRI findings of inferior olives in palatal myoclonus. *J Neurol.* 1989;236:115–116. doi:10.1007/BF00314408

Yoo YM, Lee CJ, Lee U, Kim YJ. Mitochondrial DNA in patients with essential tremor. *Neurosci Lett.* 2008;434(1):29–34. doi:10.1016/j.neulet.2008.01.023

Yu H, Sternad D, Corcos DM, Vaillancourt DE. Role of hyperactive cerebellum and motor cortex in Parkinson's disease. *Neuroimage.* 2007;35(1):222–233. doi:10.1016/j.neuroimage .2006.11.047

·　·　·　·　·

CHAPTER 6

Quantification of Tremor

Several techniques for monitoring and analysis of tremor have been developed (Table 6.1). A quantification of tremor can be achieved with kinematic tools using various miniaturized sensors applied in robotic, industrial, and aerospace fields such as accelerometers or gyroscopes, using videos or force sensors. Wireless techniques offer new perspectives for remote tracking of tremor and become part of the emerging field of body sensor networks (BSN). The choice of relevant sensors, sensor

TABLE 6.1: Available tools for quantification and monitoring of tremor	
Clinical scales and scores	The most common scales and procedures are: —Activity of Daily Living (ADL) scales, among which Extended Nottingham ADL Scale —Clinical Tremor Rating Scale (TRS) —Unified Parkinson Disease Rating Scale (UPDRS) —Nine Hole Peg Test (9HPT) —Scoring of spirals
Electromyography	—Surface EMG electrodes (SEMG) —Needle EMG electrodes —Single motor units studies —Long-term recordings (SEMG)
Kinematics	—Accelerometers —Gyroscopes —Electromagnetic tracking systems —Flexible angular sensors —Videos —Systems combining various kinematic sensors
Haptic devices	—Force-feedback devices

packaging, type of communication, power consumption, autonomy, integration, and ergonomic aspects are important factors.

Needless to say, the appropriate placement of the sensor on the body is a key factor also. It is important to take into account that recordings need to be performed in a quiet room, free of vibrations or electrical interferences. Regular calibration of the instrumentation system should not be underestimated. Using standardized recording procedures is essential for intra- and interpatients comparisons.

6.1   CLINICAL EXAMINATION AND CLINICAL SCALES
6.1.1   Clinical Examination

Clinical examination remains a key step, although presenting the following three weaknesses: the assessment is subjective, it provides short-term information, and clinical examination is not correlated with mean tremor amplitude.

The first step in evaluating any patient with tremor is to characterize the tremor on inspection and to look for associated movement disorders. Activation tests are performed, and topographical distribution is noted. Clinical examination has to be careful and extensive, including assessment of cranial nerves, voluntary force in each limb segment, tendon/plantar reflexes, detailed sensory testing, and evaluation of posture/gait.

Neurological examination of a patient complaining and/or presenting tremor includes:

- the observation of the presence of vertical or side-to-side head movement (see Table 6.2 for differential diagnosis);
- speech evaluation: search for a concomitant dysarthria which is evaluated by repetition of standard sentences, for instance: "A mischievous spectacle in Czechoslovakia" (Trouillas et al., 1997);
- testing for oculomotor disturbances (ocular fixation, saccades, pursuit);
- the observation of abnormal movements occurring at rest;
- checking for the occurrence of tremor triggered by postural tasks or voluntary muscle contraction;
- looking for the presence of other abnormal movements (dystonia, tics);
- assessing amyotrophy or muscle hypertrophy;
- evaluating abnormalities of stance and gait.

Rest tremor is detectable while the patient is seated with the upper limbs resting in the lap (showing the dorsal side of the hands) or while the patient is lying down. For lower limbs, rest conditions can be achieved crossing one leg on the other or in supine position. Cognitive tasks, for

TABLE 6.2: Differential diagnosis of head tremor
Essential tremor
Dystonia
Cerebellar ataxias
Parkinson's disease
Myoclonus
Congenital nystagmus
Psychogenic

instance, counting down and tapping of the contalateral foot, can trigger or enhance tremor (Raethjen et al., 2008).

Postural tremor is analyzed by the following maneuvers:

- holding the upper limbs outstretched with the hands in supination, parallel to the floor;
- index-to-index test: the patients has to maintain the two index fingers medially, pointing at each other at a distance of about 1 cm. Forearms are maintained horizontally.

The followings tests are used to evaluate *kinetic tremor* (Manto, 2002):

- finger-to-nose test: patient is asked to make movements of one upper limb with the hand first resting on the thigh and then touching the nose with the index.
- finger-to-finger test: patient touches the examiner's finger, which is moved and stopped in different locations in space.
- knee–tibia test: this maneuver is executed in supine position. The patient is asked to raise one leg and place the heel on the contralateral knee, which is kept motionless. The patient slides down the tibial surface in a regular way up to the ankle. The heel is then raised again up to the resting knee.

Assessment of *writing* includes drawing Archimedes' spiral: the subject is comfortably settled in front of a table, the sheet of the paper being fixed on the table with a tape to avoid motion artefacts. The subject is asked to perform the task without timing requirements. Dominant hand is examined (Figure 6.1). The patient is also asked to write standard sentences.

(a) (b)

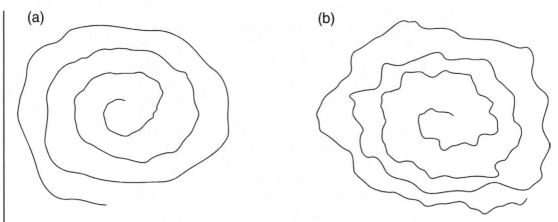

FIGURE 6.1: Drawing of a spiral in a patient with a mild essential tremor (a) and in a second patient with a more advanced disease (b).

6.1.2 Clinical Scales

Clinical scales are used to quantify the degree of severity of a clinical deficit. They systematize the information by assigning certain numbers to certain conditions.

A diagnostic measure aiming to produce quantifiable results and bound to psychometric accuracy standards should meet the following main criteria for quality:

- objectivity: independence of results from examiner;
- reliability: repeating a test application on the same person after a certain amount of time should give the same results;
- validity: a diagnostic method is valid if it actually measures the appropriate variable;
- standardization;
- comparability, economy, and usefulness.

Sensitivity (higher sensitivity means fewer false-negative cases), specificity (higher specificity means more false-positive cases), and predictive values should be taken into account (Masur, 2004).

According to *Weber's law* of psychophysics, the smallest discernible change in tremor ΔT is

$$\Delta T = K \cdot T_1,$$

where T_1 is the initial amplitude of tremor and K is Weber's constant (Gescheider, 1997). Thus, Weber's law states that the smallest discernible change in tremor will be proportional to the initial tremor amplitude. This relationship predicts that any tremor rating procedure will be a nonlinear measure of tremor amplitude. The knowledge of the relationship between tremor rating and precise measures of tremor is useful in interpreting the clinical significance of changes in ratings produced by disease or therapy. Analysis of data from 928 patients revealed a logarithmic relationship between a five-point (0–4) tremor rating procedure and tremor amplitude (Elble et al., 2006). A one-point change in TRS represents a substantial change in tremor amplitude.

The following scales are commonly used to evaluate and quantify tremor:

- *Activity of Daily Living (ADL) scale* is a questionnaire for assessing disability related to everyday functions. It is addressed to the patients themselves. Dressing, mobility, personal hygiene, eating are investigated. The measure of dependence is indicated on a defined scale. Two examples are given in the section Annexes: the *Extended ADL Scale according to Nourie and Lincoln* (Masur et al., 2004) and the *Schwab and England Activities of Daily Living.*
- *Clinical Tremor Rating Scale (TRS)* has been proposed by Fahn, Tolosa, and Marin. Tremor is evaluated through 22 items (see Annex 2).
    - The first 10 items evaluate the clinical manifestation of the tremor and its distribution in various regions of the body.
    - The other items characterize the patients' disability in daily life. Performance tests, such as handwriting and drawing, are included. The test is performed during 30 to 90 min. The range of results is 0–88 with a higher score meaning a more debilitating disease (Masur et al., 2004).
- *The Unified Parkinson Disease Rating Scale (UPDRS)* has been published in 1987 by Fahn, Elton, and the Members of the UPDRS Development Committee. It is composed of four sections (see Annex 3).
    - The first one assesses the psychological effects of the disease and the drugs used.
    - The second section features 13 subitems covering ADL.
    - In the third section (motor examination), the clinical symptoms of parkinsonism are assessed.
    - The fourth section describes the complications and side effects of drug therapies.
    - The test has a duration of 40–60 min. The range of results is 0–154 with impairment increasing from minimum to maximal points (0–5 = no impairment; 154 = maximum clinical impairment) (Masur et al., 2004).
- *Nine_Hole Peg Test (9HPT)* is an upper limb motor function test requiring a set of wooden equipment. The patient sits near a table and is asked to place pegs in holes. The examiner records the number of pegs placed in 50 s. The procedure has a duration of about 30 s in healthy persons (Mathiowetz et al., 1985).

6.1.2.1 Evaluation of Drawings. Recently, a new procedure of quantification of drawing of an Archimedes' spiral was developed (Figures 6.2 and 6.3), based on the off-line analysis of its digitized picture obtained with a commercial scanner (Miralles et al., 2006). The following assessment is performed: analysis by means of the cross-correlation coefficient with the spiral template, determination of the mean and the standard deviation of the distance between each point of the spiral drawing and the corresponding point of the spiral model, analysis of the reconstructed spiral using the Fourier transform. The experimental variables were found to be greater in the patients' group (subjects presenting action tremor, the majority of them affected with ET) compared to age-matched controls. A high linear correlation between them and the clinical score given by three

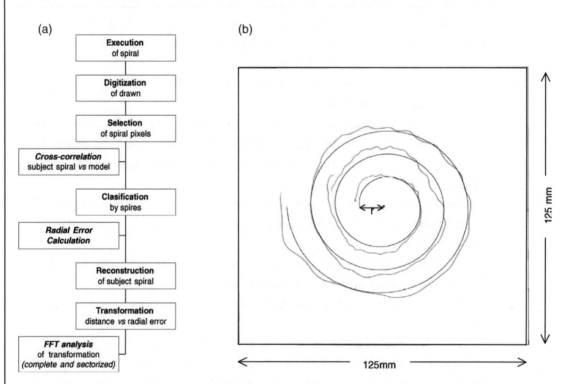

FIGURE 6.2: (a) Diagram of the different steps of the method to quantify spiral drawing. The figure shows the two parts of the procedure: the three types of numerical analysis performed on the data (left) and the progressive reconstruction of the spiral (right). (b) Digitized picture of a spiral specimen obtained from a patient with tremor showing the form and dimension of the spiral template. The initial radius (r) was 10.385 mm (i.e. 54 pixels) with an incremental change of 10.385 mm in the radius between turns. The 125 × 125 mm square that contained the spiral drawing was digitized at a resolution of 650 × 650 pixels. From Miralles et al. (2005), with permission from Elsevier.

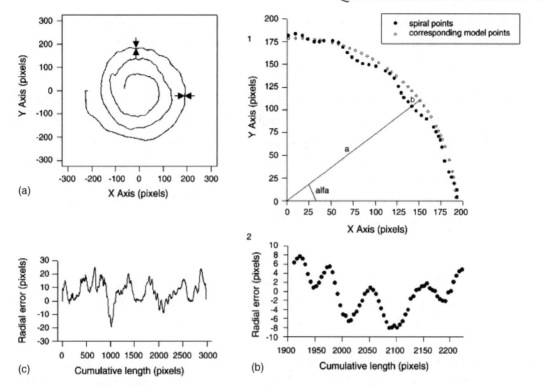

FIGURE 6.3: (a) Spiral reconstituted from the digitized picture shown in Figure 6.2b. Note the good similitude with the original image. (b, 1) Enlargement of the spiral sector located between the arrows in part (a). For the sake of clarity, only a 10% of the spiral points represented in (a) are shown. Knowing the spire to which it pertains and the angle that it forms with the x-axis (alfa), it is possible to calculate the corresponding model spiral point (white dots) for each spiral point (black dots). The radial error is signalled with the small letter b. (b, 2) Transformation of the spiral sector shown in part (b, 1). The spatial coordinates of each spiral points are substituted for another pair in which the x-value is the spiral length from the origin, and the y-value is the radial error. (c) Complete transformation of the spiral represented in part (A). From Miralles et al. (2006), with permission from Elsevier.

neurologists was also found (Spearman's rho coefficients greater than 0.7). The method is recognized to classify the spirals better than human raters.

6.1.2.2 Correlation of Postural Tremor with Quantified Tests. Alusi and colleagues found out that right arm postural tremor scores correlate with right arm Finger-Tapping Test and 9HPT scores ($p < 0.005$). A good correlation of postural tremor scores and patient perceived disability—as quantified by the ADL questionnaire—was also found. Tremor scores from spiral drawings of both dominant and nondominant hands and dominant handwriting had a high correlation with the

9HPT. Tremor scores from the nondominant hand spirals correlated less with the tremor ADL because most of the items on the scale are usually performed by the dominant hand (Alusi et al., 2000).

6.1.2.3 Vocal Tremor Scoring System. The Vocal Tremor Scoring System (VTSS) has been developed recently on the basis of a list of flexible fiber optic findings from patients with essential tremor of voice (ETV) and a review of multiple patient examinations of ETV. The final items include tremor of the palate, the base of tongue, the pharyngeal walls, the larynx (global), supraglottis, and true vocal folds. Such a standardized flexible nasolaryngoscopic examination protocol evaluates vocal tremor in a consistent fashion and could serve to predict which subpopulation of patients with ETV would benefit most from intralaryngeal muscle botulin toxin therapy (Bovè et al., 2006; see also Chapter 8).

6.2   ELECTROMYOGRAPHY AND MOTOR UNIT STUDIES

Surface EMG (SEMG) and needle EMG studies are commonly used to assess muscle discharges. Surface electrodes are arranged approximately 2 cm apart over the muscle. They record electrical potentials generated by the muscle fibers (see Chapter 3). Multiple muscles can be recorded simultaneously. Single motor unit discharges can be also studied to determine the firing rate of individual motor units, the synchronization of units within one muscle, the synchrony and coherence of tremor in homologous muscle groups (Bain, 1995). In healthy subjects, a moderate contraction is associated with motor unit discharges of about 9–16 Hz. Different diseases are associated with distinct durations of EMG bursts detected with surface sensors (Figure 6.4). In a given sensorimotor context, the patterns of motoneuronal discharges might constitute a "motoneuronal signature" which takes into account the size of the recruited motor units, the timing of their respective firings, their firing rates, and their modulation (Manto et al., 2008).

For intramuscular studies, pairs of intramuscular signals are usually recorded from a very small multichannel sterile electrode implanted in the muscle of interest. Decomposition of the myoelectric signals identifies individual motor unit action potentials (MUAPs). Firing statistics are extracted, based upon the template-matching approach (Calder et al., 2008), resolving the spatio-temporal superposition of motor unit potentials, tracking and identifying action potentials from different motor units. Motor unit potentials are identified on the basis of shape and firing time characteristics. Baseline needs to be stable enough. Sampling frequency is usually >5 kHz. The following parameters are extracted for each given task:

- the firing times of each motor unit
- the firing rates

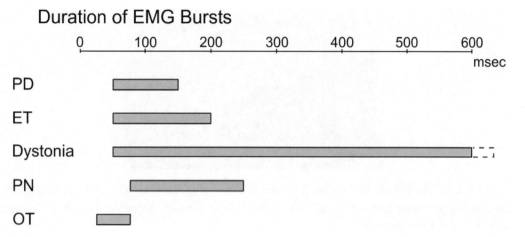

FIGURE 6.4: Duration of electromyographic (EMG) bursts in forearm muscles in various forms of tremor. PD: Parkinson's disease, ET: essential tremor, PN: peripheral neuropathy (mainly paraproteinemic neuropathy, chronic inflammatory demyelinating polyneuropathy CIDP, and hereditary neuropathies such as Charcot–Marie–Tooth disease), OT: orthostatic tremor. In some cases of dystonic tremor, the EMG bursts may last up to 800 ms.

- the recruitment threshold and the time of derecruitment
- the time of peak discharge
- the cross-correlation of the firing rates of concomitant motor units
- the synchronization histogram
- the presence of fibrillations/positive sharp waves (these are markers of denervation; fibrillation potentials have a duration <5 ms, an amplitude below 1 mV, and a discharge frequency between 1 and 50 Hz)
- the presence of fasciculations (polyphasic electrical potentials related to large motor units, see also Chapter 5: amyotrophic lateral sclerosis ALS)
- an index of instability of the motor unit potentials
- the presence of satellite potentials (markers of reinnervation)
- an index of the jitter and blocks of satellite components.

The *Tremor Coherence Analyzer* (TCA) is an electronic device developed for wireless monitoring of physiological variables such as EMG potentials or EEG traces (Figure 6.5). This technique uses surface EMG signals for tremor analysis. The portable system calculates instantaneously the coherence function between these signals, allowing the determination of linear dependencies between two signals (Brunetti et al., 2004; see Section 5).

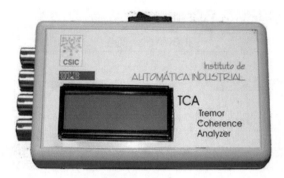

FIGURE 6.5: The Tremor Coherence Analyzer (TCA) is a wireless portable tool calculating instantaneously the coherence between 2 and 4 EMG channels to assess the behavior of the tremor generator and to detect sudden oscillators.

Long-term EMG has proven to be a valid and reliable method for the quantification of pathological tremors. Recent works indicate that these recordings allow a rater-independent classification of parkinsonian versus ET (Breit et al., 2008). A comparison of the technique of EMG with accelerometers and gyroscopes is given in Table 6.3.

The *entrainment test*—a quantified electrophysiological entrainment test performed using accelerometer or surface EMG tremor signals—has been proposed as a sensitive and specific procedure to distinguish psychogenic tremor from dystonic and other organic tremors (McAuley and Rothwell, 2004). The subject is asked to tap at a given frequency with one upper limb, while keeping the contralateral upper limb outstreched. The basic concept is the inability of patients with hysterical tremor to generate voluntary tapping oscillations independently of ongoing tremor oscillation (see Chapter 4). Normal subjects are not able to "mimic" organic tremor: a given tapping frequency will influence strongly the contralateral limb.

Simultaneous use of EMG recorded, during functional magnetic resonance imaging (fMRI) in patients with movement disorders, is complementary to conventional fMRI and may facilitate studies of hyperkinetic movement disorders, including tremor (van Rootselaar et al., 2007).

Surface and needle EMG studies are better interpreted when results of nerve conduction studies, somatosensory-evoked potentials, transcranial magnetic stimulation are available. Additional tests such as resetting of tremor, reciprocal inhibition (H reflex), stretch reflexes, ballistic movements may provide useful information in selected cases.

6.3 KINEMATIC STUDIES

Electronic positional sensors have found various clinical applications including the analysis of general physical activity, gait, posture, trunk, and upper limb movement. Tables 6.4 and 6.5 summarize

TABLE 6.3: Comparison of three sensors			
TYPE OF SENSOR			
	GYROSCOPE	**ACCELEROMETER**	**ELECTROMYOGRAM**
Measure	Measures rotational motion	Measures linear motion and gravity concurrently	Measures electrical potentials associated with muscle activity
Influence of gravity	Not influenced	Signal magnitude corrupted by gravity	Not influenced by gravity
Accuration of frequency information	Down to DC (zero frequency)		Frequency information appropriate
Accuration of magnitude information	Down to DC (zero frequency)		Magnitude less reproducible Contact resistance is a significant variable
Integration required to obtain position signal	Only a single integration is needed to obtain angular displacement	Second integration required to obtain displacement	
Signal-to-noise ratio	High signal-to-noise ratio	Low signal-to-noise ratio Gravity component	Low signal-to-noise ratio
Dynamic range	High	High	Low
Electrical contact with subject	No intimate electrical contact with subject	No intimate electrical contact with subject\	Intimate electrical contact with subject required
Size and influence on motion	Solid state gyros (with no spinning wheels) do not influence motion of subject being measured	Piezoelectric accelerometers can be very small with no influence on motion; servo accelerometers much larger	Small sensing elements available, with little influence on motion

TABLE 6.4: Summary of applications of different sensors

MAIN APPLICATIONS	SENSOR				
	ACCELEROMETER	GYROSCOPE	ELECTROMAGNETIC	FLEXIBLE ANGULAR	SENSING FABRICS
Gait analysis	√	√	X	√	X
Posture and trunk movement analysis	√	√	√	√	√
Physical activity analysis	√	X	X	X	X

Adapted from Wong et al. (2007).

TABLE 6.5: Summary of different parameters usually measured using the positional sensors

TYPE OF SENSORS	PARAMETERS MEASURED
Accelerometer	Orientation of body segment
	Acceleration of lower limbs
	Velocity and translations of lower limbs
	Angle of lower limbs
	Acceleration and angle of upper limb movements
	Frequency of upper limb movements
	Acceleration of trunk
	Step and cycle time of walking
	Metabolic energy expenditure
	Tilting angle of trunk
Gyroscope	Velocity and stride length
	Joint angle of lower limbs
	Angular velocity of trunk rotation
	Angular displacement of trunk motions

the sensors' applications and measured parameters in different studies. For historical and practical reasons, accelerometers tend to be the most commonly used (Wong et al., 2007).

New advances in microelectromechanical and wireless technologies enable inertial sensing as an alternative for motion caption. For instance, Brunetti et al. (2006) built up a wireless inertial sensor including three linear accelerometers, three gyroscopes, and three magnetometers. The use of low-mass sensors is important to decrease the low-pass filtering effect due to mass addition.

6.3.1   Accelerometry

Accelerometer (Figure 6.6) measures acceleration along the sensitive axis of the sensor based on Newton's second law (force = mass × acceleration). There are three main types of accelerometers: piezoelectric, piezoresistive, and capacitive types. The characteristics are shown in Table 6.6 (Wong et al., 2007).

Accelerometry is simple, relatively reliable, and remains a convenient technique to measure frequency and amplitude of oscillatory signals (Bain, 1995). Sensors are fixed on the skin at given

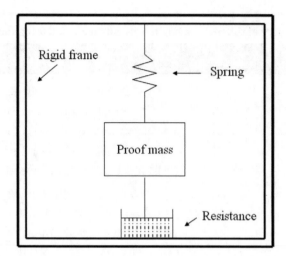

FIGURE 6.6: Basic layout of an accelerometer. In the mass–spring–damper system, the loading force drives a second-order damped harmonic oscillator, where the displacement of the proof mass relative to the rigid frame is considered. Under a constant acceleration condition, the displacement is directly proportional to the given acceleration. Modified from Wong et al. (2007).

anatomical landmarks. Various levels of support of the limbs (Figure 6.7) have been proposed (Morrison and Newell, 2000).

Accelerometers are interesting for intraoperative assessment of disabling motor symptoms such as tremor and side effects during surgery, to optimize the target position electrodes for deep brain stimulation (DBS). The intraoperative neurophysiological monitoring improves sensitivity and adds objective neurophysiological data to the clinical tests for improvement of the quality of stereotactic intervention of movement disorders (Journee et al., 2007). The reader is also referred to Chapter 8 about stereotactic surgery for tremor.

Regular calibration of accelerometers should not be underestimated.

6.3.2 Gyroscopes

A gyroscope is an angular velocity sensor. The angular orientation can then be obtained from integration of the velocity signal (Wong et al., 2007). A gyroscope is based on the measurement of the Coriolis force of vibrating devices. Coriolis force is an apparent force arising in a rotating reference frame and proportional to the angular rate of rotation. A simple model of gyroscope is shown in Figure 6.8.

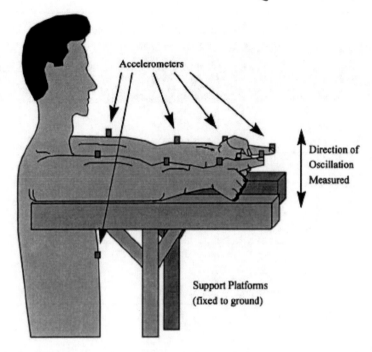

FIGURE 6.7: Schematic diagram of an experimental setup showing the position of the support and body during a condition where all segments of the upper limb—except the index fingers—were supported. From Morrison and Newell (2000), with permission from Elsevier.

There are three basic types: spinning rotor gyroscope, ring laser gyroscope, and vibrating mass gyroscope. The last one presents many advantages for portable applications because of its size, weight, lower power consumption, and cost. An internal mass is constantly vibrating inside the sensor.

6.3.3   Electromagnetic Tracking Systems
The system consists of a transmitter generating a low-frequency magnetic field detected by receivers. Signal processing difficulties may be encountered with this option in case of tremor evaluation.

6.3.4   Flexible Angular Sensor
The change of electrical output resulting from an angular change is measured. Strain gauges in goniometers produce an electrical output which is proportional to the angle. This option is clinically viable.

| PARAMETERS | PIEZOELECTRIC | PIEZORESISTIVE | | CAPACITIVE |
		SILICON	THICK FILM	
Gravitational component	No	Yes	Yes	Yes
Bandwidth	Wide	Moderate	Low	Wide
Self-generating	Yes	No	No	No
Impedance	High	Low	Low	Very high
Signal level	High	Low	Low	Moderate
Linearity	Good	Moderate	Moderate	Excellent
Static calibration (turnover)	No	Yes	Yes	Yes
Cost	High	Low	Low	High
Ruggedness	Good	Moderate	Moderate	Good
Suitable for shock	Yes	No	No	No

TABLE 6.6: Comparison of different accelerometers

Adapted from Wong et al. (2007).

6.3.5 Videos

Videotaping of patients and computerized video motion detecting systems are qualitative and quantitative methods useful to analyze motion disorders (Zhang et al., 1999). Measuring systems based on video imaging are effective in quantifying motor impairments in clinical settings (Swider, 1998). The videotaped UPDRS motor examination (see Section 6.1 above) is widely used for diagnosing PD, with some limitations when patients have mild symptoms (Louis et al., 2002). Several experimental procedures for assessment of tremor validation of new methods benefit from videotaping (Louis et al., 1998, 2000, 2001; Stacy et al., 2007). Videotaping may also be used as a teaching tool to improve the uniform application of tremor rating scales by raters with various levels of experience

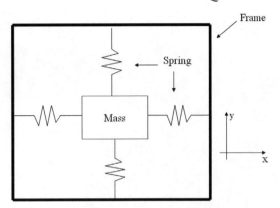

FIGURE 6.8: A simple model for analysing gyroscope behavior. A single mass is suspended by four springs in *x* and *y* directions. The frame is presumed to rotate about the *z*-axis. Assuming small-amplitude motions which will be true in resonant gyroscopes, the *x*-motion and *y*-motion can be coupled only through the Coriolis-force term and the term involving angular acceleration. The Coriolis force then induces motion in the third direction which is perpendicular both to the direction of rotation and to the driven motion. Adapted from Wong et al. (2007).

in movement disorders (Louis at al., 2001). However, analysis of videos may be time-consuming, which may become an obstacle.

6.3.6   Optoelectronic Devices

Several optoelectronic devices are commercially available. Most of them use reflective or active (light-emitting diodes LEDs) markers fixed on the skin. A calibration procedure with a reference structure may be required. Although rather expensive, their accuracy and the opportunity to perform simultaneous recordings in multiple sites are advantages.

6.4   HAPTIC DEVICES

Haptic robots and brain–computer interfaces (BCIs) are being used more and more in medicine and virtual reality. We have designed and built a new tool to investigate and monitor sensitively the functions of the wrist. This portable robotic device combines haptic technology with EMG assessment (a mechatronic myohaptic device). The system, called *wristalyzer* (Figure 6.9), allows assessing wrist motion in physiological and pathological conditions by applying loads and mechanical oscillations, taking into account the ergonomy and the angular positioning of the joints, thanks to

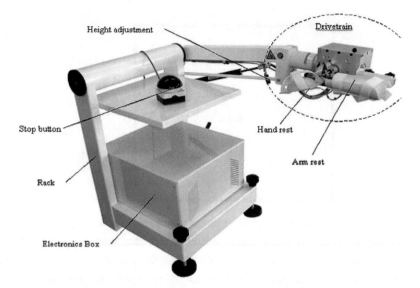

Height adjustment

Drivetrain

Stop button

Hand rest

Arm rest

Rack

Electronics Box

FIGURE 6.9: Illustration of the wristalyzer with its main components: rack with wheels, main unit, drivetrain. The mechatronic device includes a position encoder embedded in the motor casing and a torque sensor on the moving part attaching the hand flap (manipulandum) to the motor.

an adaptable manipulandum. The device characterizes the effects of damping on voluntary motion in neurological patients (Grimaldi et al., 2008).

The wristalyzer's features are the following:

- It works in a free or loaded mode for assessment of metrics of motion (normometria, dysmetria) and tremor. Angular motion, torques, and corresponding EMG activities are recorded. Spectral analysis of position, torque, and EMG trace is performed in real time.
- The device analyses the behavior of the wrist joints and the associated muscle activities during delivery of mechanical oscillations at frequencies from 0.1 to 50 Hz. Therefore, EMG activities in response to stretches can be assessed (Figure 6.10). The stretch responses are abnormal in neurological patients with a paraneoplastic cerebellar syndrome exhibiting dysmetria and kinetic tremor. The reliability of mechanically delivered oscillations has been investigated. Comparison with a conventional accelerometer has demonstrated a high accuracy in terms of reproducibility of frequency and amplitudes of oscillations (Grimaldi et al., 2007).
- The maximal voluntary contraction (MVC) can be measured.
- The device assesses automatically the impedance of the wrist (for assessment of rigidity or spasticity).

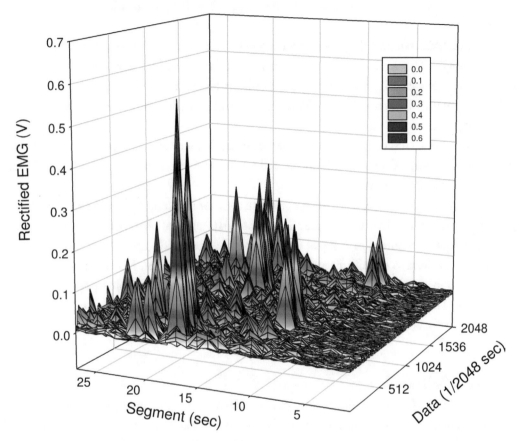

FIGURE 6.10: Effects of high-frequency oscillations (15 Hz) on EMG activity (forearm muscle) in a healthy subject. Note the bursts of EMG activities elicited by rhythmical stretches.

The wristalyzer has typical applications in acute and chronic neurological patients and can be used for rehabilitation or training purposes.

6.5   SIGNAL PROCESSING

6.5.1   Editing

Editing remains a critical step in digital processing. Data editing can be compared to a pre-analysis procedure allowing the detection and removal of degraded signals (DC offset). For tremor recordings, noise may result, for instance, from a problem which occurred during acquisition (instrumentation error, recording in a noisy environment, low frequency noise because of movement artifacts).

Many experts edit through visual inspection of recordings presented graphically. Signal-to-noise ratio (ratio of peak-to-peak signal to root mean square—RMS or quadratic mean—noise) is estimated.

We will not detail here the general principles and techniques of data filtering. Readers are referred to specialized books. Noise minimization is generally required for tremor signals. Frequency-selective filters (lowpass, highpass, bandpass, bandstop filters,or "notch" with a lower and upper cutoff) or adaptive filters may be used. Wavelet (time-scale distribution) denoising may be applied also to bandpass filter a given signal.

In case of triaxial accelerometry, the three-dimensional acceleration vector corresponds to the square root of the sum of the acceleration squared on all three axes. Removal of the 1 G (9.81 m/s^2) gravity vector can be performed.

Tremorous activity is composed of deterministic (nonrandom) and stochastic components. Signal processing is required to interpret time-series data of nonlinear systems and instances in which the frequency content of a signal provides more information than the original waveform. Most spectral estimation algorithms are devised for complete data sets. However, missing samples may occur. Nonparametric adaptative filtering-based tools are now available to deal with missing data samples. The "Gibbs phenomenon" relates to errors at points of discontinuities (a value of about 9% occurs in any region of discontinuity).

6.5.2   Fourier Transforms

Fourier transforms (FT) break down time-domain signals into constituent sinusoids of different frequencies. For periodic functions, the original waveform may be reconstructed from the sinusoidal components by application of the FT. The most used forms of FT are:

- the Fourier trigonometric series;
- the discrete FT: it is considered as a subset of the LaPlace transform;
- the fast FT (FFT): is the most popular method to perform transformation in the frequency domain. The FFT takes advantage of the symmetrical properties of sinusoidal periodic waveforms.

The spectral estimation methods allowing the calculation of the power spectra are the autocorrelation function, the FFT, and the autoregression. Units for power spectra are in power per frequency band.

Physiological waveform data generally represents truncated data segments of continuous signals. A discrete Fourier transform is often used for signals composed of data sampled at given spaced intervals, and a continuous signal may be reconstructed without information loss if the sampling frequency is greater than twice the highest frequency component in the signal (Nyquist critical frequency), to avoid aliasing. Direct application of the FT is typically modified by windowing (data windows such as Blackman/Hamming/Hanning window are applied to reduce the spectral leakage; truncated ends produce artificial side lobes or peaks in the autospectrum, see the Gibbs phenomenon above) and averaging to reduce artifacts in the power spectrum resulting from the analysis of truncated data. Indeed, spectra are averaged to reduce the variance of spectral estimates). Results of spectral densities may be gathered in a time-frequency representation (Figure 6.11).

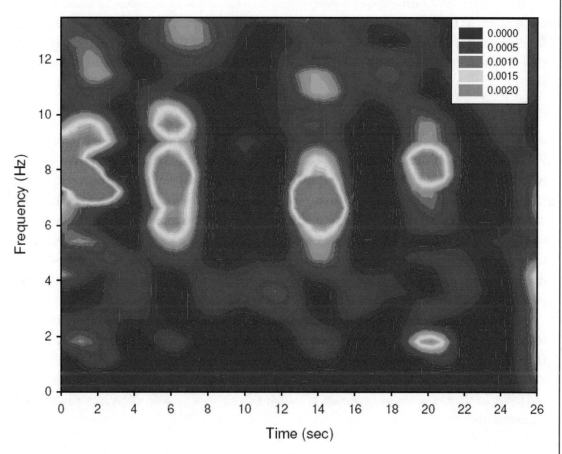

FIGURE 6.11: Time–frequency representation in a patient presenting essential tremor. Spectral densities are obtained using a Hamming window.

6.5.3 Drawbacks of FFT Analysis

There are several drawbacks of this technique. The signal is linearly decomposed as combination of sines and cosines, but most tremors are nonperiodic. The compromise between time and frequency resolution of these methods may not underline the presence of local oscillations in the signal, which might bring key information for the understanding of tremor.

The analysis of short data segments can be done using an autoregressive method of spectral analysis. The time series are first fitted to an autoregressive equation of order k. This can be very useful, for instance, in the analysis of the finger-to-nose test or the heel-to-knee test.

The Hilber analysis for tremor is a recently developed signal processing technique comprising the empirical mode decomposition (EMD) and the Hilbert spectrum (HS; Rocon et al., 2006; Figure 6.12). EMD can automatically detect and isolate tremulous movements from nonshaking movements. EMD decomposes arbitrary time-series into sets of components called intrinsic mode functions (IMFs). Subsequently, the energy of IMFs is assessed as a function of time and frequency on the HS. By contrast with the spectrogram, HS provides a windowing independent time–frequency representation allowing for the detection of events not detected by the conventional analysis.

6.5.4 Spectral Analysis Commonly Applied to Physiological Signals in Tremor

In the field of tremor, data from EMG and accelerometer recordings are treated with classical spectral analysis. Determination of spectrum shape, frequency of spectral peaks, *center frequency* (median value of the area below the power spectrum), harmonic index, as well as changes in frequency over time are some of the parameters resulting from tremor analysis. *Frequency dispersion* is the frequency width of an interval around the center of frequency that contains 66% of the total power spectrum. Frequency dispersion is used to estimate the harmonicity of oscillations. The power within a frequency range may be calculated by the integral of over the relevant frequency range (8–12 and 20–40 Hz).

These parameters allow to distinguish pathological and physiological tremor and to characterize the various types of pathological tremor. In the majority of patients with pathological tremor (PD tremor, ET, and cerebellar tremor), a single spectral peak is found (Machowska-Majchrzak et al., 2007). This peak is persistent for the whole registration period. Control subjects typically show variability in time in terms of frequency of tremor (Farkas et al., 2006). Moreover, pathological tremors show:

- significantly lower frequencies of the largest peak in the power spectrum,
- lower central frequency,

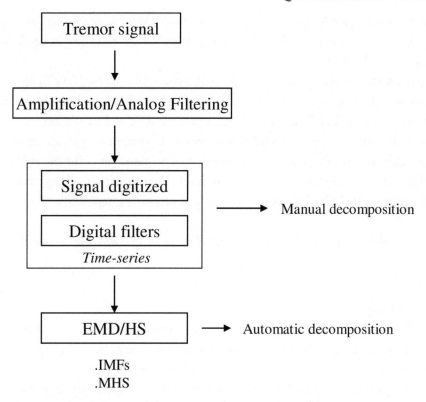

FIGURE 6.12: Block diagram showing the sequence of analysis (adapted from Rocon et al. 2006). Time-series are extracted and decomposed either manually or automatically. Tremor component and voluntary movement are demarcated. The empirical mode decomposition (EMD) provides the intrinsic mode functions (IMFs). Marginal Hilbert spectrum (MHS) $h(\omega)$ is estimated according to the following equation:

$$h(\omega) = \int_0^T H(\varpi, t)\mathrm{d}t \text{ where } \varpi(t) \text{ is the instantaneous frequency.}$$

- a significantly smaller standard deviation of the central frequency,
- a significantly higher harmonic index.

6.5.5   Wavelets

Wavelet transforms colocalize in both frequency domain and time domain and may be used effectively for nonstationary signals such as tremor. The basic idea behind wavelets is to express a signal as a linear combination of given sets of functions (wavelet transform). These are obtained by shifting and dilating a function called mother wavelet. A set of coefficients are computed. The signal is reconstructed as a linear combination of the basic functions weighted by these coefficients. Several mother wavelets are available (Gaussian, Haar, and Daubechies wavelets may be particularly useful for the assessment of EMG bursts).

Wavelet transforms are classified into discrete wavelet transforms (DWT) and continuous wavelet transforms (CWT; Figure 6.13). DWT is based on a specific subset of scale and translation values or representation grid. CWT runs over every possible scale and translation.

Because the signal in tremor has often many transient components which are interesting to isolate and analyze and because of noisy components, wavelet-based denoising can efficiently isolate activities of interest (myoelectric bursts) in many cases. Wavelet denoising has some advantages compared to conventional filtering such as smoothing, allowing a separation of the signal from noise.

Wavelet analysis can be combined with neural network analysis. This is out of the scope of this book.

6.5.6 Fuzzy Logic

In the field of the fuzzy logic (see Chapter 8), a comprehensive methodology has been developed for tremor analysis, including classic and chaotic parameters, complementing the time–frequency analysis. Teodorescu and coworkers have shown that tremor includes a significant chaotic component. The technique appears useful in diagnosis and rehabilitation. Their system includes nonlinear

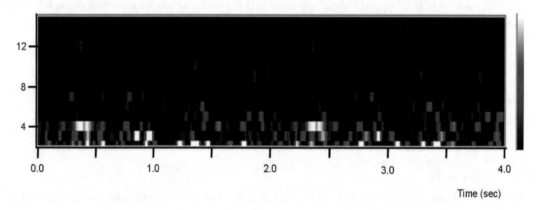

FIGURE 6.13: Continuous wavelet transform (CWT) in a postural tremor of upper limbs. Mother wavelet used: Haar filter. White corresponds to a strong positive correlation between the wavelet shape and the tremor oscillations. Black corresponds to the largest negative value.

data analysis and a fuzzy method to process the signal and to assess the type of tremor (Teodorescu et al., 2001).

6.5.7   Cross-Spectral Analysis and Coherence

Cross-spectral methods provide a powerful tool to investigate the relation between simultaneously recorded signals. These methods have been used in tremor research to study the relation between

- muscle activity (EMG) and magnetoencephalogram (MEG),
- EMG signal and electroencephalogram (EEG),
- EMG signal and other medical measurements,
- EMG signal from different muscles,
- EMG from single motor units and surface EMG.

Moreover, cross-spectral methods might be helpful in daily clinical routine as a diagnostic tool (Timmer et al., 2000).

The estimation of the *power spectrum* of a zero mean process $X_{(t)}$ is defined as the Fourier transform (FT) of the autocovariance function and is performed by a direct spectral estimation based on the FT of the measured data. The *cross-spectrum* $CS(\omega)$ of two zero mean process $X_{(t)}$ and $Y_{(t)}$, similar to the univariate case, is defined as the FT of the cross-correlation function:

$$CCF_{(t')} = (X_{(t)}Y_{(t-t')})$$

The modulus of the cross-spectrum $CS(\omega)$ normalized by the respective autospectra $Sx(\omega)$ and $Sy(\omega)$ gives the *coherence spectrum*:

$$\mathrm{Coh}(\omega) = |CS(\omega)| / \sqrt{Sx(\omega)\, Sy(\omega)}$$

The *coherence* can be understood as a measure of linear predictability. The coherence value is 1 whenever $X_{(t)}$ is obtained from $Y_{(t)}$ by a linear operator. The coherence value is equal to zero whether there is no relation or the relation between the processes is a quadratic one; interpretation of coherence does not rely on the linearity of the processes $X_{(t)}$ and $Y_{(t)}$. Simultaneously recorded signals ($X_{(t)}$ and $Y_{(t)}$) may be uncorrelated or present a coherence unequal to 1 because of:

- a nonlinear relationship
- additional influences on $X_{(t)}$ apart from $Y_{(t)}$ (other inputs to the system have not been accounted for)
- estimation bias because of misalignment
- observational noise (Timmer et al., 2000).

REFERENCES

Alusi SH, Worthington J, Glickman S, Findley LJ, Bain PG. Evaluation of three different ways of assessing tremor in multiple sclerosis. *J Neurol Neurosurg Psychiatry.* 2000;68(6):756–760. doi:10.1136/jnnp.68.6.756

Bain P. Are neurophysiological techniques useful? In: *Handbook of Tremor Disorders.* Findley LJ, Koller WC (eds). New York: Marcel Dekker, 1995.

Bové M, Daamen N, Rosen C, Wang CC, Sulica L, Gartner-Schmidt J. Development and validation of the vocal tremor scoring system. *Laryngoscope.* 2006;116(9):1662–1667. doi:10.1007/s00415-008-0712-2

Breit S, Spieker S, Schulz JB, Gasser T. Long-term EMG recordings differentiate between parkinsonian and essential tremor. *J Neurol.* 2008;255(1):103–111.

Brunetti FJ, Rocon E, Pons JL, Manto M. The tremor coherence analyzer (TCA): a portable tool to assess instantaneous inter-muscle coupling in tremor. *Conf Proc IEEE Eng Med Biol Soc.* 2004;1:61–64. doi:10.1109/IEMBS.2004.1403090

Brunetti F, Moreno JC, Ruiz AF, Rocon E, Pons JL. A new platform based on IEEE802.15.4 wireless inertial sensors for motion caption and assessment. *Conf Proc IEEE Eng Med Biol Soc.* 2006 ;1 Suppl:6497–6500.

Calder KM, Stashuk DW, McLean L. Physiological characteristics of motor units in the brachioradialis muscle across fatiguing low-level isometric contractions. *J Electromyogr Kines.* 2008;18:2–15. doi:10.1016/j.jelekin.2006.08.012

Elble RJ, Pullman SL, Matsumoto JY, Raethjen J, Deuschl G, Tintner R; Tremor Research Group. Tremor amplitude is logarithmically related to 4- and 5-point tremor rating scales. *Brain.* 2006;129(Pt 10):2660–2666.

Farkas Z, Csillik A, Szirmai I, Kamondi A. Asymmetry of tremor intensity and frequency in Parkinson's disease and essential tremor. *Parkinsonism Relat Disord.* 2006;12:49–55. doi:10.1016/j.parkreldis.2005.07.008

Gescheider GA. *Psychophysics: The Fundamentals.* Mahwah, NJ: Lawrence Erlbaum Associates, 1997.

Grimaldi G, Lammertse P, Manto M. Effects of wrist oscillations on contralateral neurological postural tremor using a new myohaptic device ('wristalyzer'). *Proceedings of the 4th IEEE-EMBS*, Cambridge, UK, 2007, pp. 44–48.

Grimaldi G, Lammertse P, Van Den Braber N, Meuleman J, Manto M. Effects of inertia and wrist oscillations on contralateral neurological postural tremor using the Wristalyzer, a new myohaptic device. *IEEE Trans Biomed Circuits Syst.* 2008 (in press).

Journee HL, Postma AA, Staal MJ. Intraoperative neurophysiological assessment of disabling symptoms in DBS surgery. *Neurophysiol Clin.* 2007;37(6):467–475. doi:10.1016/j.neucli.2007.10.006

Louis ED, Ford B, Bismuth B. Reliability between two observers using a protocol for diagnosing essential tremor. *Mov Disord.* 1998;13(2):287–293. doi:10.1002/mds.870130215

Louis ED, Barnes LF, Wendt KJ, Albert SM, Pullman SL, Yu Q, Schneier FR. Validity and test–retest reliability of a disability questionnaire for essential tremor. *Mov Disord.* 2000;15(3): 516–523. doi:10.1002/1531-8257(200005)15:3<516::AID-MDS1015>3.0.CO;2-J

Louis ED, Barnes L, Wendt KJ, Ford B, Sangiorgio M, Tabbal S, Lewis L, Kaufmann P, Moskowitz C, Comella CL, Goetz CC, Lang AE. A teaching videotape for the assessment of essential tremor. *Mov Disord.* 2001;16(1):89–93. doi:10.1002/1531-8257(200101)16:1<89:: AID-MDS1001>3.0.CO;2-L

Louis ED, Levy G, Côte LJ, Mejia H, Fahn S, Marder K. Clinical correlates of action tremor in Parkinson disease. *Arch Neurol.* 2001;58(10):1630–1634. doi:10.1001/archneur.58.10 .1630

Louis ED, Levy G, Côte LJ, Mejia H, Fahn S, Marder K. Diagnosing Parkinson's disease using videotaped neurological examinations: validity and factors that contribute to incorrect diagnoses. *Mov Disord.* 2002;17(3):513–517. doi:10.1002/mds.10119

Machowska-Majchrzak A, Pierzchała K, Pietraszek S. Analysis of selected parameters of tremor recorded by a biaxial accelerometer in patients with parkinsonian tremor, essential tremor and cerebellar tremor. *Neurol Neurochir Pol.* 2007;41(3):241–250.

Manto M. Clinical signs of cerebellar disorders. In: *The Cerebellum and Its Disorders.* Manto MU and Pandolfo M (eds). Cambridge, UK: Cambridge University Press, 2002.

Manto M, Sauvage C, Roark RM. Unifying hypothesis for the motoneuronal code in neurological disorders. *Biosci Hypoth.* 2008;2:93–99. doi:10.1016/j.bihy.2008.02.011

Masur H, Papke K. *Scale and Scores in Neurology. Quantification of Neurological deficits In Research and Practice.* Stuttgart: Thieme, 2004.

Mathiowetz V, Weber K, Kashman N, Volland G. Adult norms for the Nine Hole Peg Test of finger dexterity. *Occup Ther J Res.* 1985;5:1:24–37.

McAuley J, Rothwell J. Identification of psychogenic, dystonic, and other organic tremors by a coherence entrainment test. *Mov Disord.* 2004;19(3):253–267.

Miralles F, Tarongí S, Espino A. Quantification of the drawing of an Archimedes spiral through the analysis of its digitized picture. *J Neurosci Methods.* 2006;152(1–2):18–31.

Morrison S, Newell KM. Postural and resting tremor in the upper limb. *Clin Neurophysiol.* 2000; 111(4):651–663. doi:10.1016/S1388-2457(99)00302-8

Raethjen J, Austermann K, Witt K, Zeuner KE, Papengut F, Deuschl G. Provocation of Parkinsonian tremor. *Mov Disord.* 2008;15:1019–1023. doi:10.1002/mds.22014

Rocon E, Pons JL, Andrade AO, Nasuto SJ. Application of EMD as a novel technique for the study of tremor time series. *Conf Proc IEEE Eng Med Biol Soc.* 2006;(Suppl):6533–6536.

Stacy MA, Elble RJ, Ondo WG, Wu SC, Hulihan J; TRS study group. Assessment of interrater and intrarater reliability of the Fahn–Tolosa–Marin Tremor Rating Scale in essential tremor. *Mov Disord.* 2007;22(6):833–838.

Swider M. The application of video image processing to quantitative analysis of extremity tremor in humans. *J Neurosci Methods.* 1998;84(1–2):167–172. doi:10.1016/S0165-0270(98)00111-3

Teodorescu HL, Kandel A, Hall LO. Report of research activities in fuzzy AI and medicine at USF CSE. *Artif Intell Med.* 2001;21(1–3):177–183. doi:10.1016/S0933-3657(00)00083-X

Timmer J, Lauk M, Köster B, Hellwig B, Häußler S, Guschlbauer B, Radt V, Eichler M, Deuschl G, Lücking CH. Cross spectral analysis of tremor time series. *Int J Bifurc Chaos.* 2000;10: 2595–2610.

Trouillas P, Takayanagi T, Hallett M, Currier RD, Subramony SH, Wessel K, Bryer A, Diener HC, Massaquoi S, Gomez CM, Coutinho P, Ben Hamida M, Campanella G, Filla A, Schut L, Timann D, Honnorat J, Nighoghossian N, Manyam B. International Cooperative Ataxia Rating Scale for pharmacological assessment of the cerebellar syndrome. The Ataxia Neuropharmacology Committee of the World Federation of Neurology. *J Neurol Sci.* 1997;145(2):205–211. doi:10.1016/S0022-510X(96)00231-6

van Rootselaar AF, Maurits NM, Renken R, Koelman JH, Hoogduin JM, Leenders KL, Tijssen MA. Simultaneous EMG-functional MRI recordings can directly relate hyperkinetic movements to brain activity. *Hum Brain Mapp.* 2007. doi:10.1002/hbm.20477

Wong WY, Wong MS, Lo KH. Clinical applications of sensors for human posture and movement analysis: a review. *Prosthet Orthot Int.* 2007;31(1):62–75. doi:10.1080/03093640600983949

Zhang T, Wei G, Yan Z, Ding M, Li C, Ding H, Xu S. Quantitative assessment of Parkinson's disease deficits. *Chin Med J (Engl).* 1999;112(9):812–815.

CHAPTER 7

Mechanisms of Tremor

In this chapter, we review the pathophysiology of the various forms of tremor. Principal mechanisms and presumed generators are summarized in Table 7.1. Tremors are rarely because of a single factor.

7.1   BLOOD STUDIES

Secondary causes of tremor may be identified by laboratory workup. Tremor is a classic sign of hypoglycemia. Hyperthyroidism may induce tremor, as well as high levels of toxic agents, such as medications (see Chapter 4), metals (manganese, lead, methylmercury, copper, iron overload), etc. (Sadek et al., 2003). Copper overload is associated with Wilson's disease (characterized by corneal deposits, overload of copper in liver, low ceruplasmine level in blood, and increased 24-h cupruria) (Benito-León and Louis, 2006).

A high level of monoclonal IgM in the blood may be found in cases exhibiting a postural tremor. Tremor is associated with monoclonal gammopathy of undetermined significance (MGUS) or with a chronic inflammatory demyelinating neuropathy CIDP (Bain et al., 1996). It is worthy to note that tremor can be the presenting symptom in paraproteinemic neuropathies, but some patients will exhibit tremor several years after disease onset. Peripheral neuropathy may be associated with benign IgG and IgA paraproteinemia, in the absence of antimyelin glycoprotein activity (anti-MAG).

Cerebellar syndrome and tremor may also have an autoimmune origin. Anti-Yo, anti-Hu, anti-Tr and other antineuronal antibodies (Shams'ili et al., 2003), antiglutamic acid decarboxylase antibodies (Vianello et al., 2003), and anti-GM1 ganglioside antibodies (Zappia et al., 2002) should be looked for in case of cerebellar syndrome. Patients presenting anti-Tr antibodies-associated Hodgkin's lymphoma may present myorhythmia (Wiener et al., 2003)

Genetic diseases should not be overlooked.

7.2   BRAIN IMAGING

7.2.1   Computerized Tomography

Atrophy detected by computerized tomography (CT scan) and changes in signal intensity observed on *magnetic resonance* (*MRI*) are well-known in patients with kinetic tremor (Fukuhara et al., 1994). Lesions may involve the brainstem or the cerebellothalamocortical projections. Lesions in the white

TABLE 7.1: Principal mechanisms and generators of tremor	
TYPE OF TREMOR	**MECHANISMS/GENERATORS**
Physiological tremor	• related to the mechanical properties of the oscillating limb • central component attributed to spinal interneuronal systems, subcortical oscillators, or to cortical rhythms
Enhanced physiological tremor	• enhancement of the central component of physiological tremor • a single brainstem source or bilateral oscillators closely linked at or below this level
Essential tremor	• abnormal GABAergic system (see Chapter 3) • abnormal cerebellar afferents from the inferior olive/abnormal bilateral overactivity of cerebellar connections • involvement of the thalamic nuclei • subcortical and subcorticocortical circuits, i.e., cerebellar–thalamic–cortical pathway
Rest tremor and PD	• multiple oscillatory circuits operating on similar frequencies • corticosubthalamopallidothalamic loop • external pallidum and subthalamic nucleus acting as a pacemaker • abnormal synchronization within the striatopallidothalamic pathway • abnormal coupling in a cerebellodiencephalic–cortical loop • functional disequilibrium between GABAergic and DAergic influences (>DA in the caudoventral parts of the GPi)
Kinetic tremor	• cerebellar dysfunction • abnormality of the adaptive cerebellar afferent inflow to motor cortex
Orthostatic tremor	• bilateral overactivity of cerebellar connections • oscillator located in the brainstem (inferior olive) and influenced by cerebral cortex, basal ganglia, and cerebellum

matter with or without basal ganglia involvement are observed in vascular parkinsonism (Rao et al., 2006).

7.2.2 Positron Emission Tomography

Positron emission tomography (PET) imaging investigates in vivo (by measuring regional cerebral flow rCBF and glucose/oxygen/amino acids consumption) neurochemical, hemodynamic, or metabolic processes that underlie movement disorders. PET and specific tracers to measure *dopamine transporter* (DAT) binding can provide an accurate and highly sensitive measure of degeneration in the dopamine system in PD (Antonini et al., 2001; see also Chapter 5).

7.2.3 Magnetoencephalography

Oscillations are a prominent feature of macroscopic human sensorimotor cortical activity as recorded noninvasively with electroencephalography (EEG) and magnetoencephalography (MEG). The advent of whole-scalp MEG systems allows a rapid noninvasive recording from the entire cortex and an accurate localization of neural sources (Schnitzler and al., 2000).

7.2.4 Transcranial Magnetic Stimulation

Transcranial magnetic stimulation (TMS) studies in combination with MRI have been performed mainly for the unraveling of mechanisms of sensorimotor plasticity. Many TMS abnormalities are seen in the different diseases, concurring to show that motor cortical areas and their projections are the main target of the basal ganglia dysfunction in movement disorders (Cantello, 2002).

7.3 CONTRIBUTION OF ANIMAL MODELS IN THE IDENTIFICATION OF THE CAUSE OF TREMOR

Animal models, especially in rodents and monkeys, are widely used in the field of tremor pathogenesis. Drugs under development are first assessed in these models. Figure 7.1 summarizes the sites of action of pharmacological agents used to generate animal models of tremor.

7.4 PHYSIOLOGICAL TREMOR

Although the main determinant of PT is related to the mechanical properties of the oscillating limb, an additional central component coexists. Measurement of PT in the hand and finger in a group of 117 normal subjects aged between 20 and 94 years showed a significant reduction of the tremor frequency by adding inertia (Raethjen et al., 2000). A negative correlation with hand volume was found. The significant EMG peak (50–80% of the recordings) was unaffected by mechanical

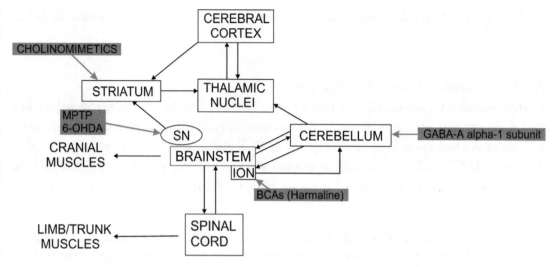

FIGURE 7.1: Sites of action of pharmacological agents used to generate animal models of tremor, especially in rodents and monkeys. Systemic administration of cholinomimetics induces a general tremor by acting on the muscarinic receptors of striatal neurons. MPTP (1-methyl-4-phenyl-1,2,3,6-tetrahydropyridine) is highly toxic for dopaminergic neurons in the substantia nigra (SN). 6-OHDA (6-hydroxydopamine) destroys cathecholaminergic fibers, including nigrostriatal projections. Beta-carboline alkaloids (BCAs) such as harmaline, harmine, and ibogaine produce an action tremor via a facilitatory effect on rhythmic activity (enhancement of electrotonic coupling) of the inferior olivary nucleus (ION) projecting to the cerebellum. Genetic mutants may exhibit a phenotype with tremor. GABA-A receptor alpha-1 subunit knockout mice exhibit a high-frequency postural and kinetic tremor. Adapted from Miwa (2007).

changes in the periphery (Figure 7.2). A significant EMG-EMG coherence was observed in about one-third of the subjects.

The independent central component in the 6- to 15-Hz band has been attributed to spinal interneuronal systems or subcortical oscillators. Recently, the role of cortical rhythms has been underlined. The findings of a 6- to 15-Hz coherence between cortex and EMG have been interpreted as a corticomuscular transmission of the oscillation rather than a peripheral feedback to the cortex (Raethjen et al., 2002).

Recent works highlight that PT occurring while the subject extends the middle finger against a spring has two load-independent components and a load-dependent component (Takanokura and Sakamoto, 2005). The stretch–reflex system causes the load-dependent component, whose frequency is determined by the mechanical property of the elastic load. The load-independent components are likely produced by the supraspinal system.

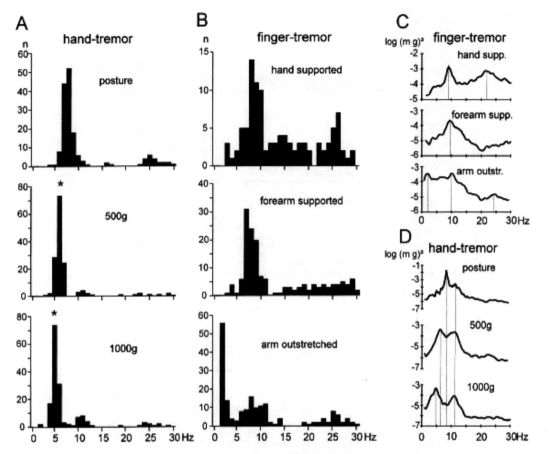

FIGURE 7.2: Distributions of accelerometric peak frequencies under different recording conditions in physiological tremor. The frequencies are displayed representatively for the right side, but the results of the other side were essentially the same. (a) The main frequency band of the hand in the 6- to 12-Hz band clearly dropped under added weight (Wilcoxon test: posture 500 g, $P < 0.001$; 500–1,000 g, $P < 0.001$). There was a small amount of remaining peaks in the 10-Hz range under weight load of the hand. (b) Finger tremor showed three distinct frequency bands (2–4, 6–12, and 15–30 Hz), which strongly depend on the arm position. (c) Examples of coincident occurrence of the different peaks within the same spectra. (d) Example of an additional hand tremor peak around 10 Hz which is not influenced by weight. From Raethjen et al. (2000), with permission from Elsevier.

The main difference in PT magnitudes between the dominant and nondominant side of right-handed individuals consists of an approximately 30% greater amplitude of fluctuations in acceleration for the nondominant hand. Central modulation of neural activity has been proposed to explain this phenomenon (Bilodeau et al., 2008).

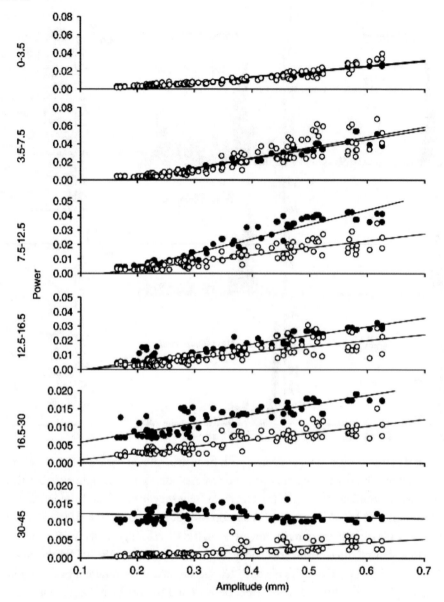

FIGURE 7.3: From top to bottom: in the ordinate axis of each plot, the sum of power found in postsurgical tremor (white dots) and matched controls for each frequency bands. Frequency bands are indicated on the far left. Abscisses represent the tremor amplitude placed in increasing order. Solid lines represent the second-order regression. Note that within the 0- to 3.5-, 3.5- to 7.5-, and 12.5- to 16.5-Hz bands, the power increased concomitantly in both groups as higher 5-s epochs amplitudes were compared. For the 7.5- to 12.5-Hz band, the power increased mostly for controls. As for the 16.5- to 30- and 30- to 45-Hz bands, there was also clearly more power for the controls. From Duval et al. (2005), with permission from Elsevier.

The cortical correlates of PT may be involved in linking different cortical motor centers and might therefore play a role in cortical motor planning (Raethjen et al., 2004).

In a comparison between post-thalatomy parkinsonian tremor and physiological tremor, Duval and colleagues found that the high-frequency component of physiological tremor failed to emerge in patients which underwent thalatomy (Figure 7.3). Therefore, they suggested that the thalamus should be considered as an important component of the generation and/or propagation of high-frequency components of physiological tremor (Duval et al., 2005).

7.5   ENHANCED PHYSIOLOGICAL TREMOR

A typical example of enhancement of postural tremor occurs in subjects treated with amitryptiline (a tricyclic antidepressant drug, see Chapter 4). The proposed mechanism is an enhancement of the central component of physiological tremor (Raethjen et al., 2001).

Transcranial magnetic stimulation (TMS) studies of long-latency reflexes (LLR) and cross-spectral analysis were performed by Köster et al. (1998) in subjects with the syndrome of persistent mirror movements (PMM), characterized by abnormal bilateral corticospinal projections (Köster et al., 1998). The authors made the hypothesis that a supraspinal mechanism of EPT would result in coherent activity between both sides in PMM subjects. The authors concluded that the 8- to 12-Hz component of EPT is transmitted transcortically.

It has been proposed that a single brainstem source or bilateral oscillators closely linked at or below this level may participate in tremor genesis of EPT (O'Sullivan et al., 2002).

7.6   POSTURAL TREMOR AND ESSENTIAL TREMOR

7.6.1   Postural Tremor

Intralimb coordination analysis in healthy subjects showed that motion is influenced by a compensatory synergy organized around the wrist and shoulder joints. This compensatory synergy reduces the coordination of the four degrees of freedom (arm links) to a single degree of freedom arm control task (Morrison and Newell, 1996). Coherence analysis of intersegment tremor relation has confirmed that the upper limbs maintain a degree of independence in both postural and rest tremor, although they exhibit the same intralimb pattern of compensatory organization.

7.6.2   Essential Tremor

Slight increases in the concentrations of glutamate and reduced levels of GABA, glycine, and serine in cerebrospinal fluid (CSF) provide a neurochemical basis for the central oscillations observed in ET and also raise the possibility that a genetically determined metabolic disorder is involved in its etiology (Mally et al., 1996). Harmaline, a beta-alkaloid derivative, induces a temporary essential-like tremor in animals via its action at the level of the olivocerebellar pathway (Wilms et al., 1999. See also Figure 7.1). The inhibition of GABA-A receptors results in enhanced electrical coupling of cerebellar afferents in the inferior olive (Stratton and Lorden, 1991). Harmaline might also act directly upon other brain structures involved in tremor genesis.

Deletion of GABA-A receptor α1 subunits determines the loss of 50% of all GABA-A receptors in the brain (Kralic et al., 2002), including the motor pathways in the brainstem, cerebellum, thalamus, and basal ganglia. Mice carrying this deletion exhibit motor incoordination that mimics human ET, although tremor frequency is higher in rodents (Kralic et al., 2005).

Another argument for the involvement of the inferior olivary nuclei and thalamus in the generation of ET comes from the significant glucose hypermetabolism of the medulla and thalami in ET patients (Hallet and Dubinsky, 1993).

The excitability of the cerebellothalamocortical pathway in ET seems normal, although recent data indicate that this might not be true at a cellular level. Resetting of tremor is feasible with motor cortex stimulation but not with cerebellar stimulation, suggesting that the abnormal oscillatory activity in ET does not originate in the cerebellar cortex itself, but that the abnormal activity is likely transmitted through the cerebellothalamocortical pathway and driven by abnormal cerebellar afferent input from areas such as the inferior olive (Pinto et al., 2003). The involvement of the thalamic neurons could be because of the fact that they generate intrinsic membrane oscillations (see also the inferior olive), behaving as oscillators or even resonators (Llinas and Paré, 1995). Thalamocortical firing patterns vary with their membrane potential. The interaction between their time-dependent cation current, the low-threshold calcium conductance, and the calcium-dependent conductance can generate oscillations from 0.5 to 4 Hz.

The case of a patient with ET who improved after a sensorimotor stroke (a small cortical infarct near the left precentral region) supports the presence of cortical or transcortical motor loops mediating ET (Kim et al., 2006). The functional changes in the sensorimotor cortex of patients with ET after stereotactic ventralis intermedius (Vim) thalamotomy (see also Chapter 8) indicate that the fundus of the central sulcus (Brodmann area 3a, see Chapter 2) is one of the key relays in the tremor circuit (Miyagishima et al., 2007). Indeed, deep brain stimulation (DBS) of the thalamic region is effective on tremor in ET, and recent studies reveal an increased efficacy of stimulation within the subthalamic area compared to the thalamus (Vim), suggesting a role of this structure (Herzog et al., 2007).

It seems plausible that tremor oscillations build up in different subcortical and subcorticocortical circuits, which temporarily entrain each other (Raethjen et al., 2007). The absence of significant coherence between the magnetoencephalogram and electromyogram at given tremor frequencies suggests that the tremor is imposed on the active muscle through descending pathways other than those originating in the primary motor cortex (Halliday et al., 2000; Figure 7.4). Coherence analysis of EMG signals in ET indicates that tremor in the right and left arm are related to independent

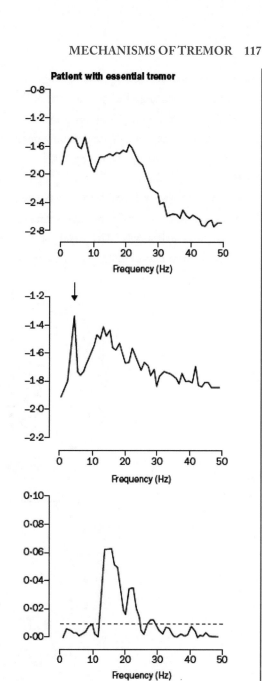

FIGURE 7.4: Spectral and coherence analysis of the magnetoencephalogram and contralateral electromyogram recorded from a healthy individual and a patient with essential tremor during voluntary maintained postural contraction of the first dorsal interosseous muscle. The horizontal dashed line indicates the level of the 95% CI for coherence. Values of coherence above this level are significant. Arrow indicates peak tremor frequency. From Halliday et al. (2000), with permission from Elsevier.

central oscillators. Hellwig and colleagues have proposed that central oscillators in the right and left brain are not entirely independent of each other and that they may dynamically synchronize, presumably by interhemispheric coupling via the corpus callosum (Hellwig et al., 2003; Figure 7.5).

Arguments for a primary involvement of cerebellum in ET genesis do exist. Proton magnetic resonance spectroscopic imaging (MRS) shows a decrease in cerebellar *N*-acetylaspartate/total creatine (NAA/tCr) ratio in ET, signifying cerebellar neuronal dysfunction or even predegeneration. A correlation between blood harmane concentration and brain NAA/tCR is found for the cerebellar cortex, but not in central cerebellar white matter, cerebellar vermis, thalamus, or basal ganglia (Louis et al., 2007). Although MRI voxel-based morphometry showed a lack of a consistent decrease in gray and white matter density in patients with ET (arguing against a progressive neurodegenerative

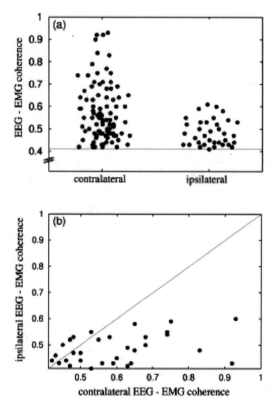

FIGURE 7.5: (a) All contralateral and all ipsilateral EEG–EMG coherences encountered in the study. The horizontal line at a coherence of about 0.4 indicates the level of significance. (b) For all cases of bilateral EEG–EMG coherence, the contralateral EEG–EMG coherences are plotted against the ipsilateral ones. From Hellwig et al. (2003), with permission from Elsevier.

process in ET; Daniels et al., 2006), heterogeneous and degenerative anatomopathological changes are reported in ET. Indeed, cerebellar changes and brainstem Lewy bodies distinguished 10 ET cases from 12 controls (Louis et al., 2006). A further study on 33 ET and 21 control brains confirmed the presence in ET brains of pathological changes in the cerebellum (75.8%)—probably as a consequence of an overactivity within cerebellar nuclei—with decreased numbers of Purkinje cells, presence of Lewy bodies in the brainstem (24.2%), degeneration of the dentate nucleus in two cases of ET without Lewy bodies, Purkinje cell heterotopias, and dendrite swellings (Louis et al., 2007). The reduction in Purkinje cell number in the brains of patients with ET has been demonstrated even in the absence of Lewy bodies, supporting the view that the cerebellum is anatomically, as well as functionally, abnormal in these cases of ET (Axelrad et al., 2008).

It is established that the cerebellum plays a role in the event-based timing of synchronized repetitive movements. Farkas et al. (2006) found a severe deficit of event-based rhythm generation on both sides, supporting the presumed bilateral cerebellar dysfunction in ET. Moreover, impairment of motor imagery (the process of mental representation of movements) was recently demonstrated. However, whether this is related to a genuine cerebellar or motor cortical dysfunction is still unclear (Lo et al., 2007).

In the subgroup of ET patients with predominant kinetic tremor, a relative expansion of gray matter areas involved in higher-order visuospatial processing is detectable by MRI voxel-based morphometry. Hypothetically, this is related to a long-term adaptive reorganization in the central nervous system, compensating the higher demands on the visuospatial control of skilled movements in case of trembling (Daniels et al., 2006).

7.7   REST TREMOR AND PARKINSON'S DISEASE

In a study on coherence in 22 subjects affected by Parkinson's disease (PD), no consistent pattern across patients was found, suggesting that tremor in PD is generated by multiple oscillatory circuits, which tend to operate on similar frequencies (Ben-Pazi et al., 2001; Raethjen et al., 2000). PD tremor is coupled within, but not between, limbs. The anatomy of basal ganglia loop (see Chapter 2) may explain the presence of several generators.

Three main neuronal mechanisms have been hypothesized:

- corticosubthalamopallidothalamic loop generating tremor
- pacemaker consisting of the external pallidum and the subthalamic nucleus
- an abnormal synchronization within the whole striatopallidothalamic pathway leading to a loss of segregation (Deuschl et al., 2000).

Typical PD resting tremor (4–6 Hz) is associated with strong coherence between the EMG of forearm muscles and activity in the contralateral primary motor cortex (M1) at tremor frequency but also at double tremor frequency. Tremor-related oscillatory activity within a cerebral network has been demonstrated. There is an abnormal coupling in a cerebellodiencephalic–cortical loop, cortical motor (primary motor cortex, cingulated/supplemental motor area, lateral premotor cortex), and sensory (secondary somatosensory cortex, posterior parietal cortex) areas contralateral to the tremor hand (Timmermann et al., 2003). The most striking differences between parkinsonian patients and healthy subjects imitating the resting tremor are a reduction of the coupling between primary sensorimotor cortex and a diencephalic structure—most likely the thalamus—and an enhancement of the coupling between premotor and primary sensorimotor cortex (Figure 7.6). These results indicate that the coupling of oscillatory activity within a cerebellodiencephalic–cortical loop constitutes a basic feature of physiological motor control, sustaining the hypothesis that parkinsonian resting tremor involves oscillatory cerebrocerebral coupling in a physiologically preexisting network (Pollok et al., 2004).

Animal studies suggest that neuronal oscillations are spontaneously generated within the basal ganglia system, especially from the GPe and the subthalamic nucleus (STN), but are mainly synchronized by cortical activity via the striatal inputs (see Cassim et al., 2002). Dopamine depletion causes a global enhancement of oscillations within the whole basal ganglia system, particularly in the GP–STN network. Interestingly, acute DA blockade (using high doses of haloperidol) generates synchronous oscillatory activity across the basal ganglia neuron populations, and prolonged DA blockade causes phase lag changes in pallidostriatal synchrony (Burkhardt et al., 2007).

The small amplitude postural tremor in PD arises from heterogeneous oscillator mechanisms. The associated increased corticomuscular coupling is related to the cortical involvement (Caviness et al., 2006).

Some authors have observed that low-frequency stimulation of the pedunculopontine region can induce tremor in the normal-behaving primate and that chronic high-frequency deep brain stimulation (DBS) of the zona incerta (which has reciprocal connections with several cortical areas, the upper brainstem, the cerebellum, and the thalamus) suppresses proximal limb tremor. This confirms the role of this area around the upper brainstem in the control of movement (Nandi et al., 2002).

Autopsy studies in PD and controls have shown that dopamine (DA) levels in the external globus pallidus (GPe) of normal brains are greater than in the internal pallidum (GPi). In PD, the mean loss of DA is marked (-82%) in GPe and moderate (-51%) in Gpi. However, DA levels are nearly normal in the ventral (rostral and caudal) GPi of PD cases with prominent tremor. There is marked loss of DA (-89%) in the caudate and severe loss (-98.4%) in the putamen in PD. The pattern of pallidal DA loss does not match the putaminal DA loss. The possible functional disequilibrium between GABAergic and DAergic influences the balance in favor of DA in the caudoventral

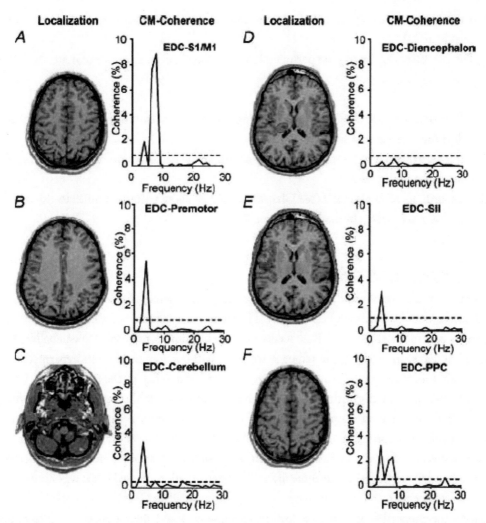

FIGURE 7.6: Localization of cerebral sources and coherence spectra to the right extensor digitorum communis (EDC) muscle in a representative subject imitating the PD resting tremor with the right hand. Note that only the S1/M1 source was localized with respect to the EDC. All other sources were detected with reference to the S1/M1 source. Coherent activity was found in the (a) contralateral primary sensorimotor cortex, (b) premotor cortex, (c) ipsilateral cerebellum, (d) diencephalon, (e) secondary somatosensory cortex, and (f) posterior parietal cortex. All sources were significantly coherent to the right EDC with the exception of the thalamic source where, in this subject, discernible spectral coherence peaks at tremor, and twice the tremor frequency failed to reach significance. Note that the strength of source localization is color coded. Red represents stronger coherent activation. The dotted lines indicate 95% confidence level of coherence. From Pollok et al. (2004), with permission from Blackwell Publishing.

parts of the Gpi, which may contribute to rest tremor in tremor-dominant and classic PD cases (Rajput et al., 2008).

The arguments against a pure peripheral mechanism generating rest tremor are the following (Llinas and Paré, 1995):

1. Rest tremor is not abolished by sectioning the dorsal roots, indicating that it does not reflect the sole action a spinal reflex loop.
2. It is very difficult to reset rest tremor by a mechanical perturbation, and the phase shift lasts for only a few cycles.
3. Recordings of Ia afferents (see Chapters 1 and 2) show patterns similar to the one found during a voluntary alternating movement.

7.8   KINETIC TREMOR

Reversible cerebellar lesions by cooling the dentate nuclei in monkeys cause ataxic movements with or without tremor (Flament and Hore, 1986). The tremor is attenuated or abolished by addition of mass and occurs in the absence of visual feedback, showing similar properties to cerebellar terminal tremor in humans. In reaching the target, movement without tremor is often hypermetric (overshoot), while movement with tremor shows an increased variability of the end position (Flament and Hore, 1986). Both discontinuities and tremor in movements during cerebellar dysfunction could result from impaired stretch reflexes and disorganized servo assistance mechanisms, both partly involving transcortical pathways (Hore and Falment, 1986). A detailed analysis of firings of neurons revealed that the population of neurons discharging strongly in relation to cerebellar tremor responded markedly and reciprocally to limb perturbation. No evidence was found that the 3- to 4-Hz cerebellar tremor is driven by a purely central oscillator (Hore and Flament, 1988). By comparing the properties of tremor following movements about the elbow with those following isometric contractions, Flament and Hore concluded that (1) a cerebellar tremor can occur in an isometric situation, and (2) movement of a given joint is required for the development of a rhythmic 3- to 5-Hz cerebellar intention tremor (Flament and Hore, 1988).

Concurrent cognitive tasks and visual feedback increase the kinetic tremor in humans. Trains of repetitive transcranial magnetic stimulation (at 10–30 Hz and intensities of 90–120% motor threshold) over the motor cortex while normal subjects make either voluntary wrist movements or during a postural task (maintenance of a fixed position) result in a tremor clinically very similar to cerebellar tremors (Topka et al., 1999). It seems that repetitive TMS may cause a terminal tremor by interfering with the adaptive cerebellar afferent inflow to the motor cortex.

Delayed second agonist EMG burst and delayed antagonist EMG burst associated with ballistic movements has been reported in advanced cases of ET during performance of ballistic movements, arguing again for a cerebellar dysfunction in case of kinetic tremor of ET (Koster et al., 2002).

7.9   ORTHOSTATIC TREMOR

Orthostatic tremor (OT) was originally classified as a variant of ET (FitzGerald and Jankovic, 1991). Several features such as a high degree of synchrony between different muscles, the lack or poor response to β-blocking agents (such as propranolol) and ethanol, a negative family history, and the higher frequencies of oscillations, suggest that OT and ET tremor are different disorders, although some overlap might exist (McManis et al., 1993; Boroojerdi et al., 1999).

A bilateral overactivity of cerebellar connections is considered as a common feature of several tremulous disorders, as demonstrated by the finding of abnormal bilateral cerebellar activation in orthostatic, essential, and writing tremors (Wills et al., 1996). OT secondary to pontine lesions may arise from dysfunction of the cerebellum or related pontine structures (Benito-León et al., 1997). The resetting by transcranial magnetic stimulation in OT associated with cerebellar cortical atrophy emphasizes the role of motor cortex in the genesis of this subtype of OT (Manto et al., 1999). Involvement of cranial muscles and failure of peripheral loading to modify tremor frequency argue strongly for a central origin (Köster et al., 1999). The existence of either a unique oscillator or a potent linking mechanism between key structures might explain the high intermuscular coherence. The fact that high-frequency discharges appear during isometric muscle contraction (irrespective of body posture) points toward a disorder of isometric force control rather than a disorder of stance regulation (Boroojerdi et al., 1999).

Sharott and colleagues showed that a 16-Hz activity may appear in healthy subjects, when unsteadiness is triggered by vestibular galvanic stimulation (Sharott et al., 2003). They suggested that the core abnormality in primary OT is an exaggerated sense of unsteadiness when standing still, which elicits activity from a 16-Hz oscillator normally engaged in postural responses. The importance of postural set is supported by the occurrence of EMG modification during the tilt test (patient tilted to different positions) but nor during the relief (patient gradually lifted) maneuver (Spiegel et al., 2006). Comparison of sway parameters in OT patients, controls, and patients with proprioceptive loss shows that one of the important components is a disruption of the normal generation or processing of proprioceptive signals (Bacsi et al., 2005). Piboolnurak and colleagues have proposed that the oscillator is most likely located in the brainstem and is influenced by the cerebral cortex, basal ganglia, and cerebellum (Piboolnurak et al., 2005).

7.10   OTHER TREMORS

7.10.1 Multiple Sclerosis

The predominance of action tremors in patients with multiple sclerosis (MS) points to a role for cerebellar structures or the cerebellothalamocortical pathway. Although the rarity of rest tremor in this demyelinating disease argues against an involvement of the basal ganglia, the tremors observed in MS might be generated by the basal ganglia. DBS of the nucleus ventralis oralis posterior (VOP) of the thalamus, which is the basal ganglia output nucleus of the thalamus, is effective (see also Chapter 8; Koch et al., 2007).

7.10.2 Psychogenic Tremor

Study of coherence between EMG signals from right and left arms (with the hands outstretched against gravity) shows that two different pathogenic mechanisms may play a role in psychogenic tremor: voluntary produced oscillations and an involuntary mechanism like clonus or enhanced physiological tremor (Raethjen et al., 2004). Evaluation of psychogenic tremor, while the subject carries out simultaneously two tasks, reveals the phenomenon of dual-task interference (Kumru et al., 2007; see also Chapter 6). The coexistence of psychogenic tremor with a tremor because of an anatomical lesion in the nervous system is not exceptional (comorbidity).

7.10.3 Dystonic Tremor

Reciprocal inhibition and other inhibitory reflex pathways at various levels (cerebral cortex, basal ganglia, spinal cord) seem to be reduced in dystonic patients (Deuschl et al., 2001; Deuschl, 2003). Indeed, excessive co-contraction and abnormalities in the time course of reciprocal inhibition between antagonist groups of muscles are considered as key features of some types of dystonia. Reduced speed of movement is often attributed to involuntary activation of antagonist muscles about a joint. Reciprocal inhibition studies include the Renshaw circuit, Ib inhibitory circuits, Ia inhibitory interneurons, and presynaptic inhibition.

Dystonic–postural tremor of the upper extremity may occur after a contralateral anterior thalamic infarct (Cho and Samkoff, 2000). The pathogenesis of thalamic tremor and dystonia is not precisely known. The main hypotheses are:

- interruption of afferents to the posterolateral thalamus from the brainstem and cerebellum (Miwa et al., 1996)
- destruction of thalamic neurons
- abnormal volleying of the circuitry between the thalamus and the cerebral cortex
- damage to the thalamic reticular nucleus (Cho and Samkoff, 2000).

7.10.4 Primary Writing Tremor

Studies using TMS showed shortened silent periods during maximal voluntary contraction (MVC) on both sides. This indicates that central inhibitory mechanisms are affected and strengthens the hypothesis that primary writing tremor is not a variant of focal task-specific dystonia but rather a separate nosological entity (Ljubisavljevic et al., 2006). In writer's cramp, there is also evidence of cortical motor reorganization (Byrnes et al., 2005).

Comparison between cortical and spinal excitability in patients with primary writing tremor (PWT) and patients with ET and writer's cramp suggests that the pathophysiology of PWT is distinct (Modugno et al., 2002). Nevertheless, PWT, dystonia, and ET are conditions which share pathophysiological hallmarks. Positron emission tomography reveals that both essential and writing tremors are associated with abnormal bilateral overactivity of cerebellar connections (Wills et al., 1995).

Studies on cerebral activation pattern during writing showed that cortex activation includes the contralateral premotor area (area 6), the ipsilateral prefrontal area (inferior frontal gyrus; areas 10, 44, and 47), the parietal lobule (area 40) bilaterally (>contralateral), and the cerebellum bilaterally (with a predominance ipsilaterally; Berg et al., 2000).

7.10.5 Complex Regional Pain Syndrome

Physiopathological mechanisms underlying the movement disorders associated with this condition are not yet clarified. Peripheral versus central origin as well as organic versus psychogenic origin have been considered. The following possibilities are investigated: vicious circle of reflexes involving anterior motor horn cells, direct action of substance P on motor neurons, abnormal sympathetic activity on muscle spindles, central reorganizaton mediated by retrograde axonal transport of trophic factors, diffuse activation of brain networks (Verdugo and Ochoa, 2000).

7.10.6 Holmes' Tremor

The involvement of the Guillain–Mollaret triangle (see Chapter 2) plays a key role in the genesis of midbrain tremor.

Recent neuroimaging–SPECT findings have demonstrated a dopaminergic denervation in a young case of Holmes' tremor (Guedj et al., 2007). In another case with dopamine sensitivity, brain SPECT has revealed impaired dopamine transporter activity in the basal ganglia and striatum, with diffusion tensor MRI (DTI) demonstrating missing ipsilateral tegmentofrontal connectivity and tractography showing reduced fiber connectivity of the superior and middle cerebellar peduncles on the lesioned side (Seidel et al., 2008). Taken together, these results suggest a reduced striatal dopaminergic input. Responsiveness to deep brain stimulation (DBS) of the Vim argues for a role of basal ganglia in the pathogenesis.

7.10.7 Cortical Tremor

Familial cortical myoclonic tremor with epilepsy (FCMTE) has been described in Chapter 6. The effects of low frequency repetitive transcranial magnetic stimulation (rTMS) have been recently studied in a case of cortical tremor caused by hyperexcitability of sensorimotor cortex. A decrease of acceleration total spectral power of the index finger postural tremor and a normalization of electrophysiologic parameters after 1 Hz rTMS over premotor cortex were found (Houdayer et al., 2007; Figure 7.7). Coherence analysis of EMG of forearm muscles and EEG of contralateral motor cortex showed strong cortico- and intermuscular coherence in the 8- to 30-Hz range, indicating a pathological cortical drive in FCMTE patients leading to tremulous movements (Van Rootselaar et al., 2006).

7.10.8 Palatal Tremor

Functional MR imaging (fMRI) reveals prominent bilateral neuronal activation in the putamen in essential palatal tremor (Haller et al., 2006; Figure 7.8). Inferior olivary hypertrophy may develop after pontine haemorrhage (lesion of the Guillain-Mollaret triangle, see Chapter 2) and may become a pacemaker for symptomatic palatal tremor (Yagura et al., 2007).

7.10.9 Tremor Following a Peripheral Nerve Injury

Pathophysiological mechanisms underlying this tremor are not fully established. Functional changes in afferent neuronal input to the spinal cord and secondary changes (central reorganization of the neuronal circuits) in the supraspinal centers are probably involved (Nobrega et al., 2002). EMG abnormalities in the absence of other findings (normal somatosensory evoked potentials, transcranial magnetic stimulation TMS, and MRI findings) and the persistence of tremor in REM and non-REM sleep in a patient with proximal right upper limb tremor (secondary to direct peripheral nerve lesion caused by thoracic surgery) suggest a peripheral generator (Costa et al., 2006).

7.10.10 Tremor in Peripheral Neuropathy

Nerve conduction velocities may be markedly reduced in paraproteinemic neuropathies. For instance, conduction velocities in the median nerves are reduced to 10–25 m/s (normal values >50 m/s). Although there is no correlation between slowing of nerve conduction and the presence of tremor, patients with the lowest velocities have usually the lowest frequencies (Smith, 1989). The fact that tremor peak frequency increases in patients recovering from severe polyneuropathy and in whom conduction velocities increase indicates that tremor frequency is linked to conduction velocities. The stretch reflex might be implicated in tremor genesis, the slowing in nerve conduction

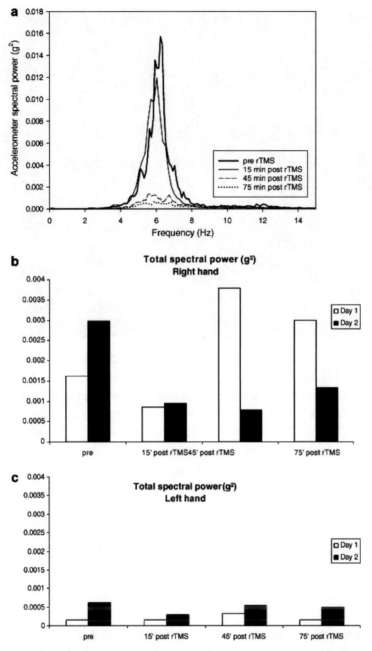

FIGURE 7.7: (a) Spectral analysis of postural tremor before and after rTMS applied over premotor cortex (PMC). (b) Accelerometer spectral power of the right hand before and three times after the first PMC stimulation. (c) Total spectral power of right (upper) and left (lower) accelerometers after the two consecutive days of PMC stimulation. From Houdayer et al. (2007), with permission from Elsevier.

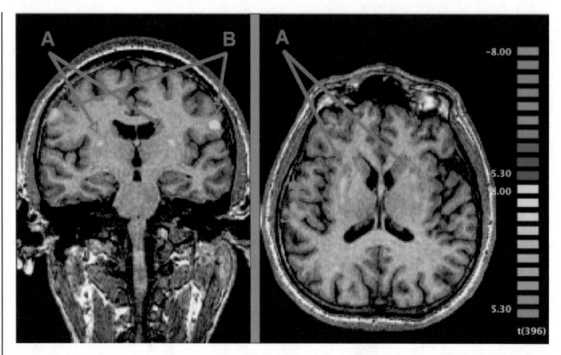

FIGURE 7.8: Neuronal activations associated with the initiation and maintenance of essential palatal tremor in a 41-year-old patient. (a) Peak activation assessed by fMRI is present in the putamen bilaterally. Additional activation is found in the (b) precentral gyrus bilaterally and right superior temporal and angular gyrus (not shown). No focal activation could be observed in the prefrontal motor cortex, the brainstem, or the cerebellum. The analysis is based on a fixed-effects general linear model with statistical threshold of $P < 0.01$ (corrected Bonferroni) and spatial threshold of 500 mm³. From Haller et al. (2006), with permission from the American Society of Neuroradiology.

velocities causing an increase in the delay of the stretch reflex loop. There is no correlation between the conduction velocity and the amplitude of oscillations.

REFERENCES

Antonini A, Moresco RM, Gobbo C, De Notaris R, Panzacchi A, Barone P, Calzetti S, Negrotti A, Pezzoli G, Fazio F. The status of dopamine nerve terminals in Parkinson's disease and essential tremor: a PET study with the tracer [11-C]FE-CIT. *Neurol Sci.* 2001;22(1): 47–48. doi:10.1007/s100720170040

Axelrad JE, Louis ED, Honig LS, Flores I, Ross GW, Pahwa R, Lyons KE, Faust PL, Vonsattel JP. Reduced Purkinje cell number in essential tremor: a postmortem study. *Arch Neurol.* 2008;65(1):101–107. doi:10.1001/archneurol.2007.8

Bacsi AM, Fung VS, Colebatch JG. Sway patterns in orthostatic tremor: impairment of postural control mechanisms. *Mov Disord.* 2005;20(11):1469–1475. doi:10.1002/mds.20600

Bain PG, Britton TC, Jenkins IH, Thompson PD, Rothwell JC, Thomas PK, Brooks DJ, Marsden CD. Tremor associated with benign IgM paraproteinaemic neuropathy. *Brain.* 1996;119(Pt 3):789–799. doi:10.1093/brain/119.3.789

Ben-Pazi H, Bergman H, Goldberg JA, Giladi N, Hansel D, Reches A, Simon ES. Synchrony of rest tremor in multiple limbs in Parkinson's disease: evidence for multiple oscillators. *J Neural Transm.* 2001;108(3):287–296. doi:10.1007/s007020170074

Benito-León J, Rodríguez J, Ortí-Pareja M, Ayuso-Peralta L, Jiménez-Jiménez FJ, Molina JA. Symptomatic orthostatic tremor in pontine lesions. *Neurology.* 1997;49(5):1439–1441.

Benito-León J, Louis ED. Essential tremor: emerging views of a common disorder. *Nat Clin Pract Neurol.* 2006;2(12):666–678; quiz 2p following 691.

Berg D, Preibisch C, Hofmann E, Naumann M. Cerebral activation pattern in primary writing tremor. *J Neurol Neurosurg Psychiatry.* 2000;69(6):780–786. doi:10.1136/jnnp.69.6.780

Bilodeau M, Bisson E, Degrâce D, Després I, Johnson M. Muscle activation characteristics associated with differences in physiological tremor amplitude between the dominant and nondominant hand. *J Electromyogr Kinesiol.* 2008 (in press).

Boroojerdi B, Ferbert A, Foltys H, Kosinski CM, Noth J, Schwarz M. Evidence for a nonorthostatic origin of orthostatic tremor. *J Neurol Neurosurg Psychiatry.* 1999;66(3):284–288.

Burkhardt JM, Constantinidis C, Anstrom KK, Roberts DC, Woodward DJ. Synchronous oscillations and phase reorganization in the basal ganglia during akinesia induced by high-dose haloperidol. *Eur J Neurosci.* 2007;26(7):1912–1924. doi:10.1111/j.1460-9568.2007.05813.x

Byrnes ML, Mastaglia FL, Walters SE, Archer SA, Thickbroom GW. Primary writing tremor: motor cortex reorganisation and disinhibition. *J Clin Neurosci.* 2005;12(1):102–104. doi:10.1016/j.jocn.2004.08.004

Cantello R. Applications of transcranial magnetic stimulation in movement disorders. *J Clin Neurophysiol.* 2002;19(4):272–293. doi:10.1097/00004691-200208000-00003

Cassim F, Labyt E, Devos D, Defebvre L, Destée A, Derambure P. Relationship between oscillations in the basal ganglia and synchronization of cortical activity. *Epileptic Disord.* 2002;4(3):S31–S45.

Caviness JN, Shill HA, Sabbagh MN, Evidente VG, Hernandez JL, Adler CH. Corticomuscular coherence is increased in the small postural tremor of Parkinson's disease. *Mov Disord.* 2006;21:492–499. doi:10.1002/mds.20743

Cho C, Samkoff LM. A lesion of the anterior thalamus producing dystonic tremor of the hand. *Arch Neurol.* 2000;Sep;57(9):1353–1355. doi:10.1001/archneur.57.9.1353

Costa J, Henriques R, Barroso C, Ferreira J, Atalaia A, de Carvalho M. Upper limb tremor induced by peripheral nerve injury. *Neurology.* 2006;67(10):1884–1886. doi:10.1212/01.wnl.0000244437.31413.2b

Daniels C, Peller M, Wolff S, Alfke K, Witt K, Gaser C, Jansen O, Siebner HR, Deuschl G. Voxel-based morphometry shows no decreases in cerebellar gray matter volume in essential tremor. *Neurology.* 2006;67(8):1452–1456. doi:10.1212/01.wnl.0000240130.94408.99

Deuschl G, Raethjen J, Baron R, Lindemann M, Wilms H, Krack P. The pathophysiology of parkinsonian tremor: a review. *J Neurol.* 2000;247(Suppl 5):V33–V48. doi:10.1007/PL00007781

Deuschl G, Raethjen J, Lindemann M, Krack P. The pathophysiology of tremor. *Muscle Nerve.* 2001;24(6):716–735.

Deuschl G. Dystonic tremor. *Rev Neurol (Paris).* 2003;159(10 Pt 1):900–905.

Duval C, Strafella AP, Sadikot AF. The impact of ventrolateral thalamotomy on high-frequency components of tremor. *Clin Neurophysiol.* 2005;116(6):1391–1399. doi:10.1016/j.clinph.2005.01.012

Farkas Z, Szirmai I, Kamondi A. Impaired rhythm generation in essential tremor. *Mov Disord.* 2006;21(8):1196–1199. doi:10.1002/mds.20934

FitzGerald PM, Jankovic J. Orthostatic tremor: an association with essential tremor. *Mov Disord.* 1991;6(1):60–64. doi:10.1002/mds.870060111

Flament D, Hore J. Movement and electromyographic disorders associated with cerebellar dysmetria. *J Neurophysiol.* 1986;55(6):1221–1233.

Flament D, Hore J. Comparison of cerebellar intention tremor under isotonic and isometric conditions. *Brain Res.* 1988;439(1–2):179–186. doi:10.1016/0006-8993(88)91474-6

Fukuhara T, Gotoh M, Asari S, Ohmoto T. Magnetic resonance imaging of patients with intention tremor. *Comput Med Imaging Graph.* 1994;18(1):45–51. doi:10.1016/0895-6111(94)90060-4

Guedj E, Witjas T, Azulay JP, De Laforte C, Peragut JC, Mundler O. Neuroimaging findings in a case of Holmes tremor. *Clin Nucl Med.* 2007;32(2):139–140. doi:10.1097/01.rlu.0000251948.30673.4f

Haller S, Winkler DT, Gobbi C, Lyrer P, Wetzel SG, Steck AJ. Prominent activation of the putamen during essential palatal tremor: a functional MR imaging case study. *AJNR Am J Neuroradiol.* 2006;27(6):1272–1274.

Hallett M, Dubinsky RM. Glucose metabolism in the brain of patients with essential tremor. *J Neurol Sci.* 1993;114(1):45–48. doi:10.1016/0022-510X(93)90047-3

Halliday DM, Conway BA, Farmer SF, Shahani U, Russell AJ, Rosenberg JR. Coherence between low-frequency activation of the motor cortex and tremor in patients with essential tremor. *Lancet.* 2000;355(9210):1149–1153. doi:10.1016/S0140-6736(00)02064-X

Hellwig B, Schelter B, Guschlbauer B, Timmer J, Lücking CH. Dynamic synchronisation of central oscillators in essential tremor. *Clin Neurophysiol.* 2003;114(8):1462–1467. doi:10.1016/S1388-2457(03)00116-0

Herzog J, Hamel W, Wenzelburger R, Pötter M, Pinsker MO, Bartussek J, Morsnowski A, Steigerwald F, Deuschl G, Volkmann J. Kinematic analysis of thalamic versus subthalamic neurostimulation in postural and intention tremor. *Brain.* 2007;130(Pt 6):1608–1625. doi:10.1093/brain/awm077

Hore J, Flament D. Evidence that a disordered servo-like mechanism contributes to tremor in movements during cerebellar dysfunction. *J Neurophysiol.* 1986;56(1):123–136.

Hore J, Flament D. Changes in motor cortex neural discharge associated with the development of cerebellar limb ataxia. *J Neurophysiol.* 1988;60(4):1285–1302.

Houdayer E, Devanne H, Tyvaert L, Defebvre L, Derambure P, Cassim F. Low frequency repetitive transcranial magnetic stimulation over premotor cortex can improve cortical tremor. *Clin Neurophysiol.* 2007;118(7):1557–1562. doi:10.1016/j.clinph.2007.04.014

Kim JS, Park JW, Kim WJ, Kim HT, Kim YI, Lee KS. Disappearance of essential tremor after frontal cortical infarct. *Mov Disord.* 2006;21(8):1284–1285. doi:10.1002/mds.20894

Koch M, Mostert J, Heersema D, De Keyser J. Tremor in multiple sclerosis. *J Neurol.* 2007;254(2):133–145. doi:10.1007/s00415-006-0296-7

Köster B, Lauk M, Timmer J, Poersch M, Guschlbauer B, Deuschl G, Lucking CH. Involvement of cranial muscles and high intermuscular coherence in orthostatic tremor. *Ann Neurol.* 1999;45(3):384–388. doi:10.1002/1531-8249(199903)45:3<384::AID-ANA15>3.0.CO;2-J

Köster B, Deuschl G, Lauk M, Timmer J, Guschlbauer B, Lucking CH. Essential tremor and cerebellar dysfunction: abnormal ballistic movements. *J Neurol Neurosurg Psychiatry* 2002;73:400–405. doi:10.1136/jnnp.73.4.400

Kralic JE, Korpi ER, O'Buckley TK, Homanics GE, Morrow AL. Molecular and pharmacological characterization of GABA(A) receptor alpha1 subunit knockout mice. *J Pharmacol Exp Ther.* 2002;302(3):1037–1045. doi:10.1124/jpet.102.036665

Kralic JE, Criswell HE, Osterman JL, O'Buckley TK, Wilkie ME, Matthews DB, Hamre K, Breese GR, Homanics GE, Morrow AL. Genetic essential tremor in gamma-aminobutyric acidA receptor alpha1 subunit knockout mice. *J Clin Invest.* 2005;115(3):774–779.

Kumru H, Begeman M, Tolosa E, Valls-Sole J. Dual task interference in psychogenic tremor. *Mov Disord.* 2007;22(14):2077–2082. doi:10.1002/mds.21670

Ljubisavljevic M, Kacar A, Milanovic S, Svetel M, Kostic VS. Changes in cortical inhibition during task-specific contractions in primary writing tremor patients. *Mov Disord.* 2006; 21(6):855–859. doi:10.1002/mds.20807

Llinas R, Paré D. Role of intrinsic neuronal oscillations and network ensembles in the genesis of normal and pathological tremors. In: *Handbook of Tremor Disorders.* Findley LJ, Koller WC (eds), Marcel Dekker, New York. 1995.

Lo YL, Louis ED, Fook-Chong S, Tan EK. Impaired motor imagery in patients with essential tremor: a case control study. *Mov Disord.* 2007;22(4):504–508. doi:10.1002/mds.21356

Louis ED, Vonsattel JP, Honig LS, Ross GW, Lyons KE, Pahwa R. Neuropathologic findings in essential tremor. *Neurology.* 2006;66(11):1756–1759.

Louis ED, Zheng W, Mao X, Shungu DC. Blood harmane is correlated with cerebellar metabolism in essential tremor: a pilot study. *Neurology.* 2007;69(6):515–520. doi:10.1212/01.wnl.0000266663.27398.9f

Louis ED, Faust PL, Vonsattel JP, Honig LS, Rajput A, Robinson CA, Rajput A, Pahwa R, Lyons KE, Ross GW, Borden S, Moskowitz CB, Lawton A, Hernandez N. Neuropathological changes in essential tremor: 33 cases compared with 21 controls. *Brain.* 2007;130(Pt 12):3297–3307. doi:10.1093/brain/awm266

Malfait N, Sanger TD. Does dystonia always include co-contraction? A study of unconstrained reaching in children with primary and secondary dystonia. *Exp Brain Res.* 2007;176(2):206–216.

Mály J, Baranyi M, Vizi ES. Change in the concentrations of amino acids in CSF and serum of patients with essential tremor. *J Neural Transm.* 1996;103(5):555–560. doi:10.1007/BF01273153

Manto MU, Setta F, Legros B, Jacquy J, Godaux E. Resetting of orthostatic tremor associated with cerebellar cortical atrophy by transcranial magnetic stimulation. *Arch Neurol.* 1999;56(12):1497–1500. doi:10.1001/archneur.56.12.1497

McManis PG, Sharbrough FW. Orthostatic tremor: clinical and electrophysiologic characteristics. *Muscle Nerve.* 1993;16(11):1254-60. doi:10.1002/mus.880161117

Miwa H, Hatori K, Kondo T, Imai H, Mizuno Y. Thalamic tremor: case reports and implications of the tremor-generating mechanism. *Neurology.* 1996;46(1):75–79.

Miwa H. Rodent models of tremor. *Cerebellum.* 2007;6:66–72. doi:10.1080/14734220601016080

Miyagishima T, Takahashi A, Kikuchi S, Watanabe K, Hirato M, Saito N, Yoshimoto Y. Effect of ventralis intermedius thalamotomy on the area in the sensorimotor cortex activated by passive hand movements: fMR imaging study. *Stereotact Funct Neurosurg.* 2007;85(5): 225–234. doi:10.1159/000103261

Modugno N, Nakamura Y, Bestmann S, Curra A, Berardelli A, Rothwell J. Neurophysiological investigations in patients with primary writing tremor. *Mov Disord.* 2002;17(6):1336–1340. doi:10.1002/mds.10292

Morrison S, Newell KM. Inter- and intra-limb coordination in arm tremor. *Exp Brain Res.* 1996;110(3):455–464. doi:10.1007/BF00229145

Morrison S, Newell KM. Postural and resting tremor in the upper limb. *Clin Neurophysiol.* 2000;111(4):651–663. doi:10.1016/S1388-2457(99)00302-8

Nandi D, Aziz TZ, Liu X, Stein JF. Brainstem motor loops in the control of movement. *Mov Disord.* 2002;17(3):S22–S27. doi:10.1002/mds.10139

Nobrega JC, Campos CR, Limongi JC, Teixeira MJ, Lin TY. Movement disorders induced by peripheral trauma. *Arq Neuropsiquiatr.* 2002;60(1):17–20. doi:10.1590/S0004-282X2002000100004

O'Sullivan JD, Rothwell J, Lees AJ, Brown P. Bilaterally coherent tremor resembling enhanced physiological tremor: report of three cases. *Mov Disord.* 2002;17(2):387–3891. doi:10.1002/mds.10097

Piboolnurak P, Yu QP, Pullman SL. Clinical and neurophysiologic spectrum of orthostatic tremor: case series of 26 subjects. *Mov Disord.* 2005;20(11):1455–1461. doi:10.1002/mds.20588

Pinto AD, Lang AE, Chen R. The cerebellothalamocortical pathway in essential tremor. *Neurology.* 2003;60(12):1985–1987.

Pollok B, Gross J, Dirks M, Timmermann L, Schnitzler A. The cerebral oscillatory network of voluntary tremor. *J Physiol.* 2004;554(Pt 3):871–878.

Raethjen J, Lindemann M, Schmaljohann H, Wenzelburger R, Pfister G, Deuschl G. Multiple oscillators are causing parkinsonian and essential tremor. *Mov Disord.* 2000;15(1):84–94. doi:10.1002/1531-8257(200001)15:1<84::AID-MDS1014>3.0.CO;2-K

Raethjen J, Pawlas F, Lindemann M, Wenzelburger R, Deuschl G. Determinants of physiologic tremor in a large normal population. *Clin Neurophysiol.* 2000;111(10):1825–1837. doi:10.1016/S1388-2457(00)00384-9

Raethjen J, Lemke MR, Lindemann M, Wenzelburger R, Krack P, Deuschl G. Amitryptiline enhances the central component of physiological tremor. *J Neurol Neurosurg Psychiatry.* 2001;70(1):78–82.

Raethjen J, Lindemann M, Dümpelmann M, Wenzelburger R, Stolze H, Pfister G, Elger CE, Timmer J, Deuschl G. Corticomuscular coherence in the 6–15 Hz band: is the cortex involved in the generation of physiologic tremor? *Exp Brain Res.* 2002;142(1):32–40. doi:10.1007/s00221-001-0914-7

Raethjen J, Lindemann M, Morsnowski A, Dümpelmann M, Wenzelburger R, Stolze H, Fietzek U, Pfister G, Elger CE, Timmer J, Deuschl G. Is the rhythm of physiological tremor involved in cortico-cortical interactions? *Mov Disord.* 2004;19(4):458–465.

Raethjen J, Kopper F, Govindan RB, Volkmann J, Deuschl G. Two different pathogenetic mechanisms in psychogenic tremor. *Neurology.* 2004;63(5):812–815.

Raethjen J, Govindan RB, Kopper F, Muthuraman M, Deuschl G. Cortical involvement in the generation of essential tremor. *J Neurophysiol.* 2007;97(5):3219–3228. doi:10.1152/jn.00477.2006

Rajput AH, Sitte HH, Rajput AH, Fenton ME, Pifl C, Hornykiewicz O. Globus pallidus dopamine and Parkinson motor subtypes. Clinical and brain biochemical correlation. *Neurology.* 2008;15:1403–1401. doi:10.1212/01.wnl.0000285082.18969.3a

Rao SS, Hofmann LA, Shakil A. Parkinson's disease: diagnosis and treatment. *Am Fam Physician.* 2006;74(12):2046–2054.

Sadek AH, Rauch R, Schulz PE. Parkinsonism due to manganism in a welder. *Int J Toxicol.* 2003;22(5):393–401. doi:10.1080/713936672

Schnitzler A, Gross J, Timmermann L. Synchronised oscillations of the human sensorimotor cortex. *Acta Neurobiol Exp (Wars).* 2000;60(2):271–287.

Seidel S, Kasprian G, Leutmezer F, Prayer D, Auff E. Disruption of nigrostriatal and cerebello-thalamic pathways in dopamine-responsive Holmes' tremor. *J Neurol Neurosurg Psychiatry.* 2008. doi:10.1136/jnnp.2008.146324

Shams'ili S, Grefkens J, de Leeuw B, van den Bent M, Hooijkaas H, van der Holt B, Vecht C, Sillevis Smitt P. Paraneoplastic cerebellar degeneration associated with antineuronal antibodies: analysis of 50 patients. *Brain.* 2003;126(Pt 6):1409–1418. doi:10.1093/brain/awg133

Sharott A, Marsden J, Brown P. Primary orthostatic tremor is an exaggeration of a physiological response to instability. *Mov Disord.* 2003;18(2):195–199. doi:10.1002/mds.10324

Smith IS. Tremor associated with peripheral neuropathy. *Electroencephalogr Clin Neurophysiol.* 1989;72:41P.

Spiegel J, Krick C, Fuss G, Sood D, Becker G, Dillmann U. Orthostatic tremor during modification of standing. *Mov Disord.* 2006;21(2):173–178.

Stratton SE, Lorden JF. Effect of harmaline on cells of the inferior olive in the absence of tremor: differential response of genetically dystonic and harmaline-tolerant rats. *Neuroscience.* 1991;41(2–3):543–549. doi:10.1016/0306-4522(91)90347-Q

Takanokura M, Sakamoto K. Neuromuscular control of physiological tremor during elastic load. *Med Sci Monit.* 2005;11(4):CR143–CR152.

Timmermann L, Gross J, Dirks M, Volkmann J, Freund HJ, Schnitzler A. The cerebral oscillatory network of parkinsonian resting tremor. *Brain.* 2003;126(Pt 1):199–212.

Topka H, Mescheriakov S, Boose A, Kuntz R, Hertrich I, Seydel L, Dichgans J, Rothwell J. A cerebellar-like terminal and postural tremor induced in normal man by transcranial magnetic stimulation. *Brain.* 1999;122(Pt 8):1551–1562. doi:10.1093/brain/122.8.1551

van Rootselaar AF, Maurits NM, Koelman JH, van der Hoeven JH, Bour LJ, Leenders KL, Brown P, Tijssen MA. Coherence analysis differentiates between cortical myoclonic tremor and essential tremor. *Mov Disord.* 2006;21(2):215–222. doi:10.1002/mds.20703

Verdugo RJ, Ochoa JL. Abnormal movements in complex regional pain syndrome: assessment of their nature. *Muscle Nerve.* 2000;23(2):198–205.

Vianello M, Morello F, Scaravilli T, Tavolato B, Giometto B. Tremor of the mouth floor and anti-glutamic acid decarboxylase autoantibodies. *Eur J Neurol.* 2003;10(5):513–514.

Wiener V, Honnorat J, Pandolfo M, Kentos A, Manto MU. Myorhythmia associated with Hodgkin's lymphoma. *J Neurol.* 2003;250:1382–1384. doi:10.1007/s00415-003-0203-4

Wills AJ, Jenkins IH, Thompson PD, Findley LJ, Brooks DJ. A positron emission tomography study of cerebral activation associated with essential and writing tremor. *Arch Neurol.* 1995;52(3):299–305.

Wills AJ, Thompson PD, Findley LJ, Brooks DJ. A positron emission tomography study of primary orthostatic tremor. *Neurology.* 1996;46(3):747–752.

Wilms H, Sievers J, Deuschl G. Animal models of tremor. *Mov Disord.* 1999 Jul;14(4):557–571.

Yagura H, Miyai I, Hatakenaka M, Yanagihara T. Inferior olivary hypertrophy is associated with a lower functional state after pontine hemorrhage. *Cerebrovasc Dis.* 2007;24(4):369–374. doi:10.1159/000106984

Zappia M, Crescibene L, Bosco D, Arabia G, Nicoletti G, Bagalà A, Bastone L, Napoli ID, Caracciolo M, Bonavita S, Di Costanzo A, Gambardella A, Quattrone A. Anti-GM1 ganglioside antibodies in Parkinson's disease. *Acta Neurol Scand.* 2002;106(1):54–57.

· · · ·

CHAPTER 8

Treatments

Given the disabling character of tremor, treatment is an important aspect of the management. In this chapter the current treatments applied for tremor disorders are discussed. Table 8.1 summarizes the therapies.

8.1 DRUGS

The first line for treatment of tremor is oral medication. β-Blockers such as propranolol, anticholinergic medications, and levodopa are effective drugs for rest tremor. Kinetic tremor may slightly respond to β-blockers, primidone, anticholinergic medication, and alcohol in case of ET (Habib-ur-Rehman, 2000). Some ataxic patients with cerebellar cortical atrophy may exhibit a benefit with gabapentin. Tremor associated with ataxia due to vitamin E deficiency (AVED) may respond to supplementation.

TABLE 8.1: Main treatments for tremors	
Drugs	Primidone, β-blockers, benzodiazepines, antiepileptics, anticholinergics, levodopa and dopamine agonists, others
Lesioning surgery	Thalamotomy, pallidotomy, subthalamotomy
γ-Knife	Thalamotomy
Deep brain stimulation (DBS)	Main targets: thalamic ventralis intermedius nucleus (Vim), subthalamic nucleus (STN), pallidum
Transcranial magnetic stimulation	Motor cortical stimulation (under investigation)
Orthosis and prosthesis	
Rehabilitation	
Brain–machine interfaces	

TABLE 8.2: Treatment of ET		
GRADE OF SEVERITY	**DRUG SUGGESTED**	**GRADE OF EFFICACY/ SIDE EFFECTS**
Disabling only during periods of stress and anxiety	Propranolol and benzodiazepines	
Disabling	Either primidone or propranolol	
If primidone or propranolol do not provide adequate control	Primidone and propranolol in combination	Propranolol and primidone reduce limb tremor
If propranolol induces side effects	Consider other beta-adrenoceptor antagonists (such as atenolol or metoprolol)	
If combination of primidone and propranolol do not provide adequate control	The use of benzodiazepines provides some benefit	−Alprazolam is probably effective in reducing limb tremor −Clonazepam possibly reduces limb tremor
Other medications that may be helpful	−Atenolol, gabapentin (monotherapy), sotalol, and topiramate −Clozapine, nadolol, and nimodipine −Botulinum toxin in the hand muscles in case of hand tremor (risk of hand weakness)	
Others drugs	Carbonic anhydrase inhibitors (e.g. methazolamide), phenobarbital, clonidine, and mirtazapine	Questionable efficacy
Disabling head or voice tremor	Botulinum toxin injections	Breathiness, hoarseness, and swallowing difficulties may occur in attempting to treat voice tremor Injection should be performed in a specialized center
If disabling tremor persists after medication trials	Surgical options are considered	Chronic DBS and thalamotomy are usually effective

The currently available medications decrease tremor in approximately 50% of the patients. The combination of medical and surgical therapies provide benefit in up to 80% of cases. From Lyons et al. (2003), Zesiewicz et al. (2005), and Pahwa and Lyons (2003)

Table 8.2 and Figure 8.1 focus on therapeutic strategies for ET. Table 8.3 lists the medications, the doses and the potential side effects (Benito-León and Louis, 2006). Drugs are presented in detail in the text; further information about the studies quoted is given in Table 8.4.

8.1.1   Ethanol

Ethanol decreases postural essential tremor, but not parkinsonian rest tremor or the genuine cerebellar kinetic tremor (which is worsened by small doses of ethanol). Ethanol improves gait ataxia in patients with ET (Klebe et al., 2005). However, the improvement is temporary and followed by a rebound phenomenon when the alcohol effect wears off. Recently, a case of ethanol responsive tremor in a patient with MS has been reported (Hammond and Kerr, 2008).

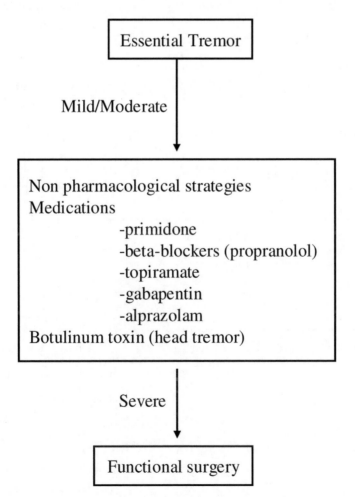

FIGURE 8.1: Therapeutic strategies of essential tremor.

TABLE 8.3: Pharmacological agents and nonpharmacological procedures for the treatment of ET

PHARMACOLOGICAL AGENTS	DOSAGE	SIDE EFFECTS/COMPLICATIONS
Alprazolam	0.125–3 mg/day	Sedation, fatigue, abuse
Atenolol	50–150 mg/day	Light-headedness, nausea, cough, dry mouth, sleepiness
Botulinum toxin A (hand tremor)	50–100 U	Hand or finger weakness, reduced grip strength, pain at injection site, hematoma
Botulinum toxin A (head tremor)	40–400 U	Neck weakness, postinjection pain
Clonazepam	0.5–6 mg/day	Drowsiness
Clozapine	6–75 mg/day	Sedation, agranulocytosis (risk of 0.8% at 1 year)
Gabapentin	1,200–1,800 mg/day	Lethargy, fatigue, decreased libido, dizziness, nervousness, shortness of breath
Levetiracetam	1,000 mg/day	Somnolence, asthenia, dizziness, headache
Nimodipine	120 mg/day	Headache, heartburn
Nadolol	120–240 mg/day	None
Olanzapine	20 mg/day	Drowsiness, sedation, weight gain, diabetes
1-Octanol	Up to 64 mg/kg	Unusual taste sensation
Primidone	Up to 750 mg/day	Sedation, drowsiness, fatigue, nausea, vomiting, ataxia, dizziness, unsteadiness, confusion, vertigo
Propranolol	60–800 mg/day	Reduced blood pressure, bradycardia, exertional dyspnea, confusion, drowsiness, headache, dizziness, impotency

PHARMACOLOGICAL AGENTS	DOSAGE	SIDE EFFECTS/COMPLICATIONS
Long-acting propranolol	80–320 mg/day	Skin eruption, dizziness
Sotalol	75–200 mg/day	Decreased alertness
Topiramate	Up to 400 mg/day	Appetite suppression, weight loss, paresthesias, attention deficit, kidney stones
Zonisamide	100–200 mg/day	Ataxia, dizziness, somnolence, agitation, anorexia
NONPHARMACO-LOGICAL AGENTS	**INTENSITY OF RESPONSE**	**POTENTIAL SIDE EFFECTS**
Chronic thalamic stimulation DBS (hand)	60% to 90% improvement in CRS (clinical rating scale)	Dysarthria, dysequilibrium, paresthesias, weakness, headache, intracranial hemorrhage, subdural hemorrhage, ischemic changes, generalized motor seizures, lead dislodgement, decreased verbal fluency
Thalamotomy	55% to 90% improvement in CRS	Hemiparesis, transient speech and motor deficits, cognitive deficit, confusion, somnolence
Gamma knife surgery	70% to 85% improvement in CRS	Transient arm weakness, numbness in the contralateral arm, dysarthria, increased action tremor, dystonia of the contralateral upper and lower limbs, choreoathetosis, delayed side effects

From Benito-León and Louis (2006) and Zesiewicz et al. (2005).

Tolerance develops with time (Habib-ur-Rehman, 2000). For obvious reasons, regular alcohol intake cannot be recommended (see also Chapter 5). Therefore, ethanol is not used in the treatment of tremor, but as a clue for the diagnosis of ET. The improvement of tremor after ethanol ingestion is likely because of a direct effect on a central oscillator (Zeuner et al., 2003). Experimental studies in rodents suggest that ethanol antagonizes a dysregulation of glutamatergic pathways in the cerebellum. In addition, it enhances GABAergic transmission.

TABLE 8.4: Details of the studies reported

REFERENCE	DRUG	MODE OF ADMINISTRATION/ DOSE	TYPE OF TREMOR/ DISEASE	DESIGN OF STUDY	N° OF PATIENTS	RESULT
–Findley et al. 1985	Primidone	–Oral	–ET of the hands and head	–Double-blind, placebo-controlled trial	–22	–Significantly superior to placebo; efficacy comparable to propranolol
–Sasso et al. 1990	Primidone	–Oral 375–750 mg/day	–ET	–Placebo-controlled	–11	–Magnitude of tremor after 3, 6, and 12 months significantly reduced
–Serrano-Dueñas 2003	Primidone	–Oral 250 vs 750 mg/day	–ET	–Double-blind study with a 1-year follow-up	–113	–Low doses equally effective
–Calzetti et al. 1992	Propranolol	–Single oral dose (120 mg) and following 2 weeks of 120 or 240 mg/day	–ET of head	–Double-blind, placebo-controlled	–9	–Significant reduction in tremor magnitude after single oral dose but not on sustained administration

Study	Drug	Dose	Condition	Design	N	Result
Koller and Herbster 1987	–Long-acting propranolol, primidone and clonazepam	–Oral 160, 250, and 4 mg/day, respectively	–PD (resting, postural, kinetic)	Double-blind crossover design	–10	–Significantly decreased of resting and postural tremor by long-acting propranolol but not by primidone or clonazepam
–Henderson et al. 1997	–Propranolol	–Oral carbimazole plus propranolol (40 mg)	–Hyperthyroidism	–Double-blind, crossover and placebo-controlled	–7	–100%
–Metzer et al. 1993	–Long-acting propranolol vs placebo	–Oral	–Drug-induced parkinsonism	–Double-blind, placebo-controlled, crossover	–6	–No significant difference
Lee et al. 2003	Arotinolol vs propranolol	10 vs 40; 20 vs 80; 30 vs 160 mg/day	–ET	Crossover, multiple dose comparative trial	165	Arotinolol as effective as propranolol
–Gates 1993	–Clonazepam and chlordiazepoxid	–0.5–2.0; 30 mg/day	–OT		–9	–8 patients responded to clonazepam and 1 to chlordiazepoxid

continued on next page

TABLE 8.4: (*cont.*)

REFERENCE	DRUG	MODE OF ADMINISTRATION/ DOSE	TYPE OF TREMOR/ DISEASE	DESIGN OF STUDY	N° OF PATIENTS	RESULT
–Gunal et al. 2000	–Alprazolam, acetazolamide, primidone	–Alprazolam 0.75 mg/day	–ET	–Double-blind, cross-over, placebo-controlled	–22	–alprazolam superior to placebo and equal to primidone; acetazolamide vs placebo: no significant differences
–Elble et al. 2007	–Levetiracetam	–5-week titration from 500 to 3,000 mg/day; max tolerated dosage for 4 weeks	–ET	–Double-blind, placebo-controlled crossover	–15	–No efficacy
–Handforth et al. 2004	–Levetiracetam	–4-week titration to 3,000 mg/day; 2 weeks of stable dose	–ET	–Placebo-controlled, double-blind, randomized crossover	–15	–No significant difference
–Ondo et al. 2004	–Levetiracetam	–Evaluation in baseline, after 2 weeks 500 mg bid; at 4 weeks 1,500 mg bid	–ET	–Open label	–10	–No efficacy

–Bushara et al 2005	–Levetiracetam	–single dose of 1,000 mg	–ET	–Double-blind, placebo-controlled	–24	–Significant reduction of hand tremor for at least 2 h
–Striano et al 2006	–Levetiracetam	–500 mg/ bid the first week up to 50 mg/kg/ day by increments of 500 mg bid each week	–MS cerebellar tremor	–Open-label	–14	–11/14 completed the trial with significant improvement
–Ondo et al. 2006	–Topiramate vs placebo	–Mean final dose of 292 mg/day	–ET	–Multicenter, double-blind, placebo-controlled, parallel-design trial	–208	–Effective
–Gatto et al. 2007	–Topiramate	–50 mg/day during 3-12 months	–ET	–Report	–3	–Effective
–Connor et al. 2008	–Topiramate	–400 mg/day or maximum tolerated dose	–ET	–Randomized, double-blind, placebo-controlled, crossover trials	–62	–Effective
Raj et al. 2006	Oxcarbazepine	–450 mg bid	ET	Report	1	Improvement

continued on next page

TABLE 8.4: (*cont.*)

REFERENCE	DRUG	MODE OF ADMINISTRATION/ DOSE	TYPE OF TREMOR/ DISEASE	DESIGN OF STUDY	N° OF PATIENTS	RESULT
Rodrigues et al. 2006	Gabapentin	600–2,700 mg/d	OT	Double-blind placebo-controlled crossover	6	Tremor amplitude reduced from 100% to 79% ± 11%
Zesiewicz et al. 2007	Pregabalin	50 mg/day up to 600 mg/day by 75 mg/day every 4 days	ET	Pilot, double-blind, placebo-controlled, randomized	22	Significant improvement
–Liou and Shih 2006	–Trihexyphenidyl	–38 mg/day	–Rubral tremor	–Report	–1	–Significant benefit
–Wills et al. 1999	–Levodopa	–	–OT		–8	–Benefit in 5/8
–Yetimalar et al. 2005	–Olanzapine and propranolol	–20 and 120 mg/day	–ET	–Randomized, crossover study	–28	–Olanzapine significantly improved tremor
–Ceravolo et al. 1999	–Clozapine	–Single dose of 12.5 mg; 39 ± 9 mg up to 50 mg in responders	–Drug-resistant ET	–Randomized, double-blind, crossover study	–15	–Single dose effective in 13/15; significant improvement with chronic treatment

Reference	Drug	Dose	Condition	Study type	N	Results
Rice et al. 1997	Ondansetron	Single dose intravenous	Cerebellar tremor	Placebo controlled, double-blind, crossover study	20	Improvement in writing and 9-hole–peg test
Gbadamosi et al. 2001	Ondansetron	Single dose intravenous	Cerebellar MS tremor	Open-label, prospective, and controlled study	14	No beneficial effects
Bier et al. 2003	Ondansetron	Oral 8 mg	Cerebellar dysfunction	Randomised, multicenter, double-blind trial	45	No beneficial effect
Jiménez-Jiménez et al. 1994	Nicardipine	Nicardipine 1 mg/kg/day and propranolol 160 mg/day; evaluation baseline and 1 month after each drug	ET	Crossover study	14	Improvement with a nonsignificantly higher efficacy of propranolol
Muenter et al. 1991	Methazolamide	Max mean dose 203 mg/day (maintenance dose 129 mg/day)	ET	–	28	21/28 mild to marked improvement
Hallett et al 1991	Isoniazid	up to 1,200 mg/day + 100 mg pyridoxine, for 4 weeks	ET and postural action tremor	Double-blind	15 (11 and 4)	Benefit in 2/11

8.1.2   Primidone

Primidone is significantly superior to placebo in reducing the magnitude of hand postural tremor. Its efficacy is similar to propranolol (Findley et al., 1985). Long-term efficacy of primidone (range of doses 375–750 mg/day) in ET has been evaluated. Tremorolytic effects last for up to 1 year or more (Sasso et al., 1990). In some studies, low doses of primidone (250 mg/day) were demonstrated to provide an equal or more effective effect than high doses (750 mg/day) in the control of ET with the subsequent advantage of fewer undesirable effects (Serrano-Dueñas, 2003). However, a progressive increase of the dose up to 750 mg/day is often required in daily practice. Low doses should be considered at the very beginning of the treatment. We start with doses in the range of 30–60 mg for 1 to 3 weeks, before increasing the dose.

8.1.3   β-Blockers

β-Adrenergic blocking drugs (mainly the nonselective blocker *propranolol* or a β2-selective blocker) have been the mainstays for the treatment of ET with limb tremor. They are probably less effective in the treatment of voice and head tremor (Habib-ur-Rehman, 2000). The efficacy of sustained propranolol on isolated or prominent essential head tremor is less predictable and satisfactory than expected on the basis of the single-dose response, compared with hand tremor (Calzetti et al., 1992).

A double-blind crossover study comparing the effects of long-acting propranolol hydrochloride (160 mg/d), primidone (250 mg at night), and clonazepam (4 mg/d) in parkinsonian patients, showed that long-acting propranolol is a useful adjuvant therapy for the parkinsonian tremor. In fact, long-acting propranolol reduced the mean amplitude of resting tremor by 70% and the mean amplitude of postural tremor by 50%, without the occurrence of noticeable side effects (Koller and Herbster, 1987). Long-acting propranolol is usually not effective to reduce drug-induced parkinsonian tremor (Metzer et al., 1993).

Interestingly, propranolol is a useful adjunct in the early treatment of both the tremor and tachycardia of hyperthyroidism (Henderson et al., 1997).

β-Blockers should be used very cautiously in the case of respiratory disease (asthma) or conduction block (heart disease), especially in the elderly.

8.1.4   Benzodiazepines

Alprazolam can be used in elderly patients with ET who do not tolerate primidone or propranolol (Gunal et al., 2000). *Clonazepam* alone or in combination with gabapentine or primidone improve orthostatic tremor (Gates, 1993) and may provide benefit in tremor associated with myoclonus.

8.1.5 Antiepileptic Drugs

Levetiracetam is not beneficial in the treatment of ET, as demonstrated by double-blind, placebo-controlled crossover trials (Elble et al., 2007; Handforth et al., 2004) and a 4-week, open label trial (Ondo et al., 2004). Levetiracetam has been proposed for the management of cerebellar tremor in multiple Sclerosis (Striano et al., 2006).

A case of ET responding to *oxcarbazepine* has been reported recently (Raj et al., 2006). *Topiramate* (up to a maximum of 400 mg/day) is effective in the treatment of moderate to severe ET. Tremor reduction is accompanied by functional improvements, such as writing or speaking (Ondo et al., 2006; Connor et al., 2008). Improvement of ET has been also reported with low doses of topiramate (50 mg/day; Gatto et al., 2007).

Gabapentin is the most effective treatment for OT. It is the first-line therapy, reducing both tremor and postural instability and improving quality of life (Rodrigues et al., 2006).

A recent study using *Pregabalin* in ET showed significant improvements in terms of accelerometry recordings and action tremor limb scores on the Fahn–Tolosa–Marin rating scale (see Annex 2). Larger studies are needed to confirm the beneficial effects (Zesievicz et al., 2007).

8.1.6 Levodopa, Dopamine Agonists, and Anticholinergic Agents

Dopamine agonists, such as *pramipexole* and *ropinirole*, are probably the most effective tremorolytic drugs among all dopaminergic treatments and should be considered in newly diagnosed Parkinson's disease (PD) patients exhibiting tremor in the absence of cognitive impairment. Improvement of tremor has also been reported with other dopamine agonists, including *pergolide* and *bromocriptine*. Dopamine agonists are also useful in advanced PD patients exhibiting a tremor refractory to levodopa and anticholinergics (Bhidayasiri, 2005). *Apomorphine (subcutaneous, puffs)* is used in advanced form of PD.

Dopaminergic and anticholinergic agents are equally effective in patients with parkinsonian tremor, but dopaminergic substances improve other parkinsonian signs also (Habib-ur-Rehman, 2000). Midbrain tremor in patients with symptomatic dystonia and mesencephalic lesions may significantly improve with levodopa (Vidailhet et al., 1999). *Levodopa* can also provide symptomatic relief in primary OT (Wills et al., 1999).

Anticholinergics, such as trihexylphenidyl, are effective, but side effects often limit their use, especially in the elderly population. Anticholinergics are not recommended in patients with cognitive decline, heart disease, or in elderly patients over 65 years of age (Bhidayasiri, 2005). In a study comparing the effects of *trihexiphenidyl, carbidopa-levodopa*, and *amantadine hydrochloride* in PD, tremor amplitude was reduced by 59% with trihexiphenidyl, 55% by carbidopa–levodopa, and 23% by amantadine (Koller, 1986). A case of successful monotherapy of midbrain tremor with high-dose trihexyphenidyl has been reported (Liou and Shih, 2006).

8.1.7 Other Drugs

Olanzapine might be effective for ET (Yetimalar et al., 2005). *Clozapine* could be considered in selected cases of resistant ET (Ceravolo et al., 1999). However, this drug requires a close follow-up.

The effects of *ondansetron* (a 5HT3-antagonist) on cerebellar dysfunction has been studied. The drug provides no benefit or minor reduction on cerebellar tremor (Rice et al., 1997; Gbadamosi et al., 2001; Bier et al., 2003).

The calcium-channel blocker *nicardipine* provides some minor benefits in ET (Jiménez-Jiménez et al.,1994).

Methazolamide might be effective in ET patients, particularly those with head and voice tremor (Muenter et al., 1991).

Jaw tremor can be successfully treated with *botulinum toxin* injections to the masseters (Gonzalez-Alegre et al., 2006).

Tremor associated with peripheral neuropathy in a context of chronic inflammatory demyelinating polyneuropathy (CIDP) may respond to steroid therapy, cytotoxic drugs, intravenous immunoglobulin therapy, and plasma exchanges (Cook et al., 1990; Dyck, 1990).

The currently available medications decrease tremor in approximately 50% of the patients. The combination of medical and surgical therapies provide benefit in up to 80% of cases. From Lyons et al. (2003), Zesiewicz et al. (2005), and Pahwa and Lyons (2003).

TABLE 8.5: Advantages of DBS compared to irreversible lesion therapy
No permanent brain lesion
Adaptable
Well tolerated in the elderly
May be performed even in patients who previously received other surgical treatments
Bilateral operations feasible

See text for references.

8.2 DEEP BRAIN STIMULATION AND SURGICAL TREATMENTS

DBS within the ventralis intermedius (Vim) nucleus is effective for tremor reduction, irrespective of the etiology. Nowadays, DBS has replaced lesioning techniques in most cases, making bilateral operations feasible, and allowing the postoperative selection of active contacts (for spatial adjustments after surgery). A great advantage is that DBS procedures replicate the effects of ablative interventions but do not require a destructive brain lesion (see Table 8.5 for the advantages of DBS; Espay et al., 2006).

A retrospective study in a total of 130 patients assessed that DBS surgery confirmed that DBS is an effective and safe method to treat PD (Seijo et al., 2007). Direct comparison of different stimulation sites in individual patients revealed that DBS in the subthalamic area is more effective in suppressing pharmacoresistant kinetic tremor than the ventrolateral thalamus proper.

The activation of A1 receptors has been proposed as a new pharmacological target to replace or improve the efficacy of DBS (Bekar et al., 2008). DBS is associated with a marked increase in the release of ATP, resulting in accumulation of its catabolic product, adenosine. Adenosine A1 receptor activation depresses excitatory transmission in the thalamus and reduces both tremor and DBS-induced side effects. Intrathalamic infusion of A1 receptor agonists directly reduce tremor.

A quality of life instrument specifically addressed to patients treated with DBS for movement disorders has been proposed (Kuehler et al., 2003).

8.2.1 Vim Stimulation

Anatomical structures possibly involved in tremor suppression include cerebellothalamic projections, the prelemniscal radiation, and the zona incerta (Hamel et al., 2007). Chronic stimulation of the thalamic ventralis intermedius (Vim) nucleus is reversible, adaptable, and well tolerated by elderly patients. Optimal electrode location for thalamic DBS in ET was found to correspond to the anterior margin of the Vim (Papavassiliou et al., 2004). Stimulator adjustments are frequently needed in ET patients, and it should be kept in mind that tremor may worsen despite a readjustment in the system (Lee and Kondziolka, 2005). Frequency of stimuli is usually between 110 and 150 Hz, with a pulse width around 60 μs.

8.2.2 STN Stimulation

STN is another effective target of DBS on tremor (Krack et al., 1998). Subthalamic stimulator implantation in a large consecutive series of patients with PD produced significant clinical improvement without mortality or major neurological morbidity (Goodman et al., 2006). STN stimulation in levodopa-intolerant patients has a great benefit in terms of reducing the level of required levodopa medication (Katayama et al., 2001).

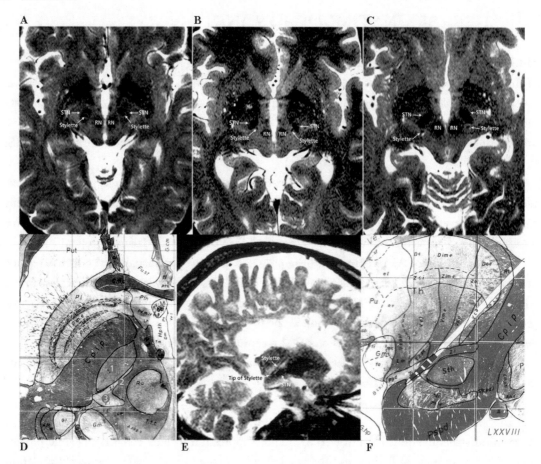

FIGURE 8.2: Intraoperative axial MRI scan showing bilateral stylettes in the **A.** STN, lateral to the anterior border of the red nucleus; **B.** medial/dorsomedial to the STN, and in the **C.** caudal zona incerta (RN: red nucleus, STN: subthalamic nucleus). **D.** Axial slice from the Schaltenbrand atlas showing the position of the three different subthalamic areas targeted. **E.** Intraoperative sagittal MRI scan showing the transfrontal trajectory of the stylette passing dorsal to the outlined STN and its tip posterior to it in the caudal zona incerta. **F.** Matching sagittal slice from the Schaltenbrand atlas with a DBS lead drawn to scale and the anatomical location of each contact on it. From Plaha et al. (2006), with permission from Oxford University Press.

The most effective contact lies dorsal/dorsomedial to the STN (region of the pallidofugal fibers and the rostral zona incerta) or at the junction between the dorsal border of the STN and the latter (Figures 8.2 and 8.3; Plaha et al., 2006). The prelemniscal radiation (RAPRL), a subthalamic bundle of fibers, has been also reported as an effective target—independent of the subthalamic nucleus—for the alleviation of tremor and rigidity in patients with PD by either lesioning or neuromodulation (Velasco et al., 2001).

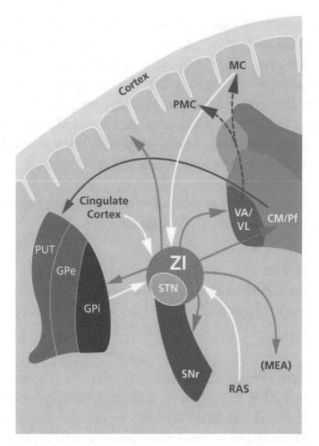

FIGURE 8.3: Schematic diagram showing the key afferent and efferent connections of the zona incerta. Afferent fibers are shown as white arrows, red arrows correspond to efferent fibers. Efferent fibers from CM/Pf to posterodorsal putamen are shown in green. PUT putamen, GPe globus pallidus externus, GPi globus pallidus internus, SNr substantia nigra reticulata, STN subthalamic nucleus, RAS reticular activating system, MEA midbrain extrapyramidal area, ZI zona incerta, CM/Pf centromedian and parafascicular nucleus of the thalamus, VA/VL ventral anterior and ventral lateral nucleus of the thalamus, MC motor cortex, PMC premotor cortex. From Plaha et al. (2006), with permission from Oxford University Press.

8.2.3   Complications of DBS

DBS is usually a safe procedure, even in patients who previously received other surgical treatments for PD, such as thalamotomy or cell grafting (Vergani et al., 2006; Fraix et al., 2005). However, serious complications leading to permanent neurologic deficit may develop after STN–DBS for advanced PD. Intracranial hemorrhage and infections are potential side effects given the surgical procedure. Intracranial hemorrhage (ICH) is considered as the most significant complication associated with the placement of stereotactic intracerebral electrodes (Sansur et al., 2007). Hypertension and the use of microelectrode recording are the main risk factors for cerebral hemorrhage. "Hardware complications," including lead migration or lead fracture are not exceptional (Lyons et al., 2004).

Reported rates of depression, cognitive impairment, mania, and behavior changes are, in principle, low when the DBS procedure has been carefully performed. However, according to some studies, there might be a higher rate of suicide in patients treated with DBS, particularly with thalamic and GPi stimulation. Therefore, a careful preoperative assessment and close postoperative psychiatric and behavioral follow-up are recommended (Burkhard et al., 2004; Appleby et al., 2007; Soulas et al., 2008).

A unique case of hyperhidrosis as a side effect of a misplaced DBS electrode near the Vim nucleus has been reported in a patient with ET (Diamond et al., 2007).

8.2.4 Lesion of the Vim

Radiofrequency lesioning in the Vim alleviates rest, postural, and action tremors affecting the distal and proximal parts of the left upper extremity in Holmes' (rubral) tremor (secondary to midbrain tumor; Kim et al., 2002).

8.2.5 Lesions of the STN or Pallidum

Unilateral radiofrequency STN lesions can be made safely and are an effective alternative to thalamotomy, pallidotomy, and unilateral STN–DBS for the treatment of asymmetrical tremor-dominant advanced PD. Combined lesioning of the dorsolateral STN and pallidofugal fibers and the zona incerta is particularly effective (Patel et al., 2003). For those PD patients whose tremor was not successfully controlled by pallidotomy, the combined pallidothalamotomy is an alternative (Llumiguano et al., 2006). Posterior globus pallidus internus pallidotomy combined with drug therapy is effective for parkinsonian leg tremor (Goto et al., 2002).

8.2.6 Complications of Lesions

PD Patients may develop aphasia (repetition, comprehension, fluency, and naming abnormalities) as a complication of thalamotomy (Bruce et al., 2004). Complications (intracranial hemorrhage, hemiparesis without hemorrhage, partial visual field deficit, speech disturbance) from stereotactic posteroventral pallidotomy (PVP) for patients with PD are relatively uncommon. However, residual symptoms may be serious (Higuchi and Iacono, 2003). Transient ataxia has been reported after nucleus ventralis intermedius thalamotomy for relief of severe parkinsonian tremor (Kimber et al., 2003). The risk of persistent morbidity following lesion therapy in patients after severe head trauma restricts the operation to selected cases with disabling tremor (Krauss et al., 1994).

8.2.7   Stimulation Versus Lesion?

Despite the usefulness and advantages of DBS (see Table 8.5) surgery in many cases of PD and ET, lesion therapy (subthalamotomy, pallidotomy, or thalamotomy) represents a viable alternative and has several advantages, including a reduced need for specialized care and close clinical follow-up, improved affordability, and a lower infection risk (Hooper et al., 2008). In ET, thalamotomy should be considered when deep brain stimulation does not succeed (Giller and Dewey, 2002). In their review of the comparison of the efficacy and safety of thalamotomy and DBS in the treatment of multiple sclerosis tremor, Yap and colleagues underline that both thalamotomy and thalamic DBS are comparable procedures for tremor suppression and that adverse effects (death, procedure-related hemorrhages, hemiparesis, dysarthria, swallowing difficulties, balance disorder) can occur with both procedures (Yap et al., 2007). In intractable MS tremor, thalamotomy has been pointed out as a more efficacious surgical treatment. However, the higher incidence of persistent neurological deficits in patients receiving lesional surgery brings support for the use of DBS as the preferred surgical strategy (Bittar et al., 2005).

8.2.8   Risk of Cognitive Deficits

A randomized trial where patients with severe drug-resistant tremor due to PD, ET, or multiple sclerosis underwent thalamotomy or thalamic stimulation showed that both thalamotomy and thalamic stimulation are associated with a minimal overall risk of cognitive deterioration and that verbal fluency decreased after left-sided thalamotomy and thalamic stimulation (Schuurman et al., 2002).

8.3   γ-KNIFE

Gamma knife (γK) radiosurgery represents a minimally invasive alternative to radiofrequency lesioning and DBS. Gamma knife thalamotomy is especially valuable for patients ineligible for radiofrequency thalamotomy or deep brain stimulation. γK radiosurgery is a viable alternative in the treatment of parkinsonian and ET in patients who are not good candidates for radiofrequency thalamotomy or thalamic DBS (Niranjan et al., 1999). Long-term follow-up of a group of 158 patients (diagnosis: PD, ET, tremor due to stroke, encephalitis, or cerebral trauma) who underwent MRI-guided radiosurgical nucleus ventralis intermedius (Vim) thalamotomy, indicates that relief of tremor is well maintained and free from radiation-induced complications (Young et al., 2000).

Gamma knife radiosurgical thalamotomy was found to be effective also for the palliative treatment of disabling multiple sclerosis intention tremor (Mathieu et al., 2007; Figure 8.4). A follow-up study showed that all patients experienced an improvement of tremor after a period of 2.5 months (median latency). One patient developed a transient hemiparesis which resolved after corticosteroid administration.

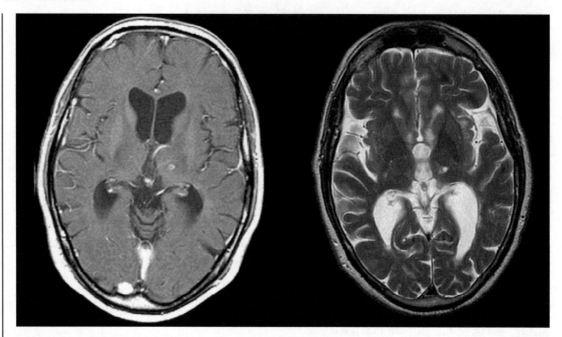

FIGURE 8.4: Brain MRI performed 2 years after γK thalamotomy. *Left*. Contrast-enhanced axial T1-weighted image demonstrating a ring-enhancing lesion in the left vim nucleus. *Right*. Axial T2-weighted image showing the lesion in the left vim nucleus with minimal changes in the surrounding brain. From Mathieu et al. (2007), with permission from Elsevier.

8.3.1   Complications

Potential complications may occur (Kondziolka et al., 2008). There is a low probability of a delayed neurological deficit. It is worthy to note that patients with ET are at risk for developing a complex, disabling movement disorder after gamma knife thalamotomy (Siderowf et al., 2001).

8.4   TRANSCRANIAL MAGNETIC STIMULATION

At this stage of development, the use of TMS has research implications. The technique is not entered yet in the field of therapeutics for tremor. Repetitive TMS (rTMS) modulates the corticospinal excitability and the effects last beyond the duration of the rTMS itself (Maeda et al., 2000). Motor cortical stimulation has been proposed as a possible therapeutic target for treatment of action tremor (Houdayer et al., 2007).

8.5   ORTHOSIS AND PROSTHESIS

In addition to medications, rehabilitation programs, and deep brain stimulation, biomechanical loading has appeared recently as a potential tremor-suppression alternative. Patients with severe

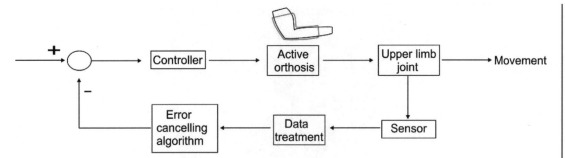

FIGURE 8.5: Tremor-suppression control system. Strategy used to control the upper limb using an active orthosis.

action tremor have uncontrollable, relatively rapid oscillatory motion superimposed on otherwise slower voluntary motor activity. The principle is that a mechanical damper produces an opposing force proportional to velocity; therefore applying damping loads to tremulous limbs should attenuate the tremulous component of movement. Aisen and colleagues have examined the effect of upper extremity damping in 10 patients with posttraumatic and multiple sclerosis cerebellar action tremor and found significant tremor reduction with the application of damping. The amounts of damping were applied by an energy-dissipating prosthesis which generated resistive viscous loads by means of computer-controlled magnetic particle brakes (Aisen et al., 1993).

A wearable orthosis applying effective dynamic force between consecutive segments of the upper limb and changing its biomechanical characteristics has been proposed as a new treatment for ET (Rocon et al., 2007). This exoskeleton acts in parallel to the affected limb and makes an online adaptive viscous control. The system makes a real-time estimation of tremor and consequently selects a cutoff frequency below the frequency of tremor (Figure 8.5). The wearable orthosis is able to detect position, rate, and acceleration of rotation of the joint by means of a chip gyroscope. The musculoskeletal system is modeled as a second-order biomechanical system exhibiting a low-pass filtering behavior.

Evaluation of this orthosis in six patients suffering from ET reveals this technique as a non-invasive alternative to medication and deep brain stimulation (Rocon et al., 2007). The robotic exoskeleton (Figure 8.6) provides a means of testing and validating nongrounded control strategies for orthotic tremor suppression (Rocon et al., 2007). Evaluation on a patient presenting a disabling ET resulted in a drastically decrease of tremor power (Figure 8.7).

Further developments are required to use a wearable orthosis in daily life. Indeed, wearable devices must present several characteristics in terms of aesthetics (appearance, size, and shape), cosmetics, and functional aspects. Functionality is related to the trade-off in terms of torque, velocity, and robustness. Design of the orthosis varies according to the shape, size, side, and functions of the limb. Current developments tend to use sensors and actuators embedded in comfortable textiles.

FIGURE 8.6: Overall view of Wotas (wearable orthosis for tremor assessment and suppression) activated by DC motors.

8.6 BRAIN–MACHINE INTERFACES

Brain–computer interfaces (BCIs) enable users to control devices using electroencephalographic (EEG) signals from the scalp or with single-neuron activity from inside the brain. BCIs aim to control limbs as smoothly as possible. Leuthardt and colleagues pointed out that electrocorticographic (ECoG) activity recorded from the surface of the brain allows controlling a one-dimensional computer cursor rapidly and accurately (Leuthardt et al., 2004). Therefore, they suggested that an ECoG-based BCI could be suggested for people with severe motor disabilities as a nonmuscular communication and control option that is more powerful than EEG-based BCIs. This BCI is potentially more stable and less traumatic than BCIs based upon electrodes penetrating the brain. The option of BCI to control tremor is under investigation.

8.7 ASSISTIVE THERAPY

A mouse computer is available for tremor sufferers, filtering the shaking component in movement. A digital smoothing filter is placed between the mouse motion sensors and the operating system software computing the coordinates of the cursor.

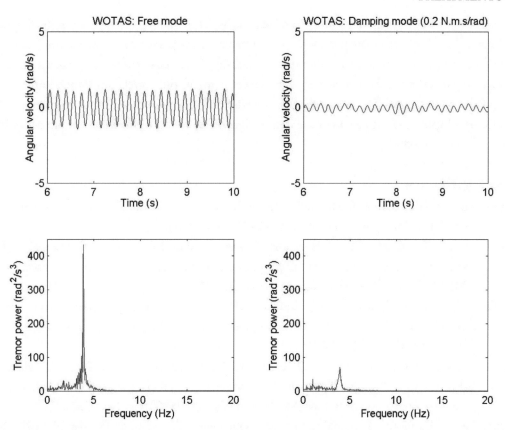

FIGURE 8.7: Oscillations of the elbow and the associated power spectral density (PSD), with the motor in a free mode (left panels) and providing viscosity of 0.2 N m s/rad (right panels) in one ET patient. Note the strong reduction in the PSD when viscosity is applied. Oscillations are expressed in rad/s and PSD is expressed in rad^2/s^3. From Manto et al. (2007), with permission from IPO Publishing Ltd.

Portable haptic devices allow to implement so-called gravity wells around selected regions of the screen.

8.8 FUZZY LOGICS

Fuzzy logic is an emerging field for tremor research and therapy (see also Chapter 6). A system for rehabilitation based on fuzzy assessment of fused features of the tremor movements has been proposed (Teodorescu et al., 2001). In this method, both classic and nonlinear indices are fused into a few, easy to represent, and grasp fuzzy measures. The purpose is the creation of a feedback to help patients to become aware of the characteristics of the tremor of their limbs and help them in controlling their tremor. One of the main requirements is the ease of use to understand and learn.

A representation of the tremor by images and sounds has been built up on the basis of the linguistic and fuzzy valuation of the main parameters of the tremor. Such use is justified by its simplicity (easy to understand by the patient). The rules used to globally characterize the tremor take into account tremor amplitude, main frequency, frequency power content, the correlation dimension, and the irregularity of the signal. The rules establish the relations between the pattern in the space of the tremor signal space and the classes of the tremor signal. The results of the inference are defuzzified and used in the feedback.

REFERENCES

Aisen ML, Arnold A, Baiges I, Maxwell S, Rosen M. The effect of mechanical damping loads on disabling action tremor. *Neurology.* 1993;43(7):1346–1350.

Alusi SH, Aziz TZ, Glickman S, Jahanshahi M, Stein JF, Bain PG. Stereotactic lesional surgery for the treatment of tremor in multiple sclerosis: a prospective case-controlled study. *Brain.* 2001;124(Pt 8):1576–1589. doi:10.1093/brain/124.8.1576

Appleby BS, Duggan PS, Regenberg A, Rabins PV. Psychiatric and neuropsychiatric adverse events associated with deep brain stimulation: a meta-analysis of ten years' experience. *Mov Disord.* 2007;22(12):1722–1728. doi:10.1002/mds.21551

Bekar L, Libionka W, Tian GF, Xu Q, Torres A, Wang X, Lovatt D, Williams E, Takano T, Schnermann J, Bakos R, Nedergaard M. Adenosine is crucial for deep brain stimulation-mediated attenuation of tremor. *Nat Med.* 2008;14(1):75–80. doi:10.1038/nm1693

Benito-León J, Louis ED. Essential tremor: emerging views of a common disorder. Nat *Clin Pract Neurol.* 2006;2(12):666–678. doi:10.1038/ncpneuro0347

Bhidayasiri R. Differential diagnosis of common tremor syndromes. *Postgrad Med J.* 2005;81:756–762. doi:10.1136/pgmj.2005.032979

Bier JC, Dethy S, Hildebrand J, Jacquy J, Manto M, Martin JJ, Seeldrayers P. Effects of the oral form of ondansetron on cerebellar dysfunction. A multi-center double-blind study. *J Neurol.* 2003;250(6):693–697. doi:10.1007/s00415-003-1061-9

Bittar RG, Hyam J, Nandi D, Wang S, Liu X, Joint C, Bain PG, Gregory R, Stein J, Aziz TZ. Thalamotomy versus thalamic stimulation for multiple sclerosis tremor. *J Clin Neurosci.* 2005;12(6):638–642. doi:10.1016/j.jocn.2004.09.008

Burkhard PR, Vingerhoets FJ, Berney A, Bogousslavsky J, Villemure JG, Ghika J. Suicide after successful deep brain stimulation for movement disorders. *Neurology.* 2004;63(11):2170–2172.

Bushara KO, Malik T, Exconde RE. The effect of levetiracetam on essential tremor. *Neurology.* 2005;64(6):1078–1080.

Bruce BB, Foote KD, Rosenbek J, Sapienza C, Romrell J, Crucian G, Okun MS. Aphasia and thalamotomy: important issues. *Stereotact Funct Neurosurg.* 2004;82(4):186–190. doi:10.1159/000082207

Calzetti S, Sasso E, Negrotti A, Baratti M, Fava R. Effect of propranolol in head tremor: quantitative study following single-dose and sustained drug administration. *Clin Neuropharmacol.* 1992;15(6):470–476. doi:10.1097/00002826-199212000-00004

Ceravolo R, Salvetti S, Piccini P, Lucetti C, Gambaccini G, Bonuccelli U. Acute and chronic effects of clozapine in essential tremor. *Mov Disord.* 1999;14(3):468–472. doi:10.1002/1531-8257(199905)14:3<468::AID-MDS1013>3.0.CO;2-M

Connor GS, Edwards K, Tarsy D. Topiramate in essential tremor: findings from double-blind, placebo-controlled, crossover trials. *Clin Neuropharmacol.* 2008;31(2):97–103.

Cook D, Dalakas M, Galdi A, Biondi D, Porter H. High-dose intravenous immunoglobulin in the treatment of demyelinating neuropathy associated with monoclonal gammapathy. *Neurology.* 1990;40:212–214.

Diamond A, Kenney C, Almaguer M, Jankovic J. Hyperhidrosis due to deep brain stimulation in a patient with essential tremor. Case report. *J Neurosurg.* 2007;107(5):1036–1038.

Dyck PJ. Intravenous immunoglobulin in chronic inflammatory demyelinating polyradiculoneuropathy and in neuropathy associated with IgM monoclonal gammapathy of unknown significance. *Neurology.* 1990;40:327–328.

Elble RJ, Lyons KE, Pahwa R. Levetiracetam is not effective for essential tremor. *Clin Neuropharmacol.* 2007;30(6):350–356.

Espay AJ, Mandybur GT, Revilla FJ. Surgical treatment of movement disorders. *Clin Geriatr Med.* 2006;22(4):813–825, vi. doi:10.1016/j.cger.2006.06.002

Findley LJ, Cleeves L, Calzetti S. Primidone in essential tremor of the hands and head: a double blind controlled clinical study. *J Neurol Neurosurg Psychiatry.* 1985;48(9):911–915.

Fraix V, Pollak P, Moro E, Chabardes S, Xie J, Ardouin C, Benabid AL. Subthalamic nucleus stimulation in tremor dominant parkinsonian patients with previous thalamic surgery. *J Neurol Neurosurg Psychiatry.* 2005;76(2):246–248. doi:10.1136/jnnp.2003.022707

Gates PC. Orthostatic tremor (shaky legs syndrome). *Clin Exp Neurol.* 1993;30:66–71.

Gatto EM, Roca MC, Raina G, Micheli F. Low doses of topiramate are effective in essential tremor: a report of three cases. *Clin Neuropharmacol.* 2003;26(6):294–296. doi:10.1097/00002826-200311000-00006

Gbadamosi J, Buhmann C, Moench A, Heesen C. Failure of ondansetron in treating cerebellar tremor in MS patients—an open-label pilot study. *Acta Neurol Scand.* 2001;104(5):308–311. doi:10.1034/j.1600-0404.2001.00075.x

Giller CA, Dewey RB Jr. Ventralis intermedius thalamotomy can succeed when ventralis intermedius thalamic stimulation fails: report of 2 cases for tremor. *Stereotact Funct Neurosurg.* 2002;79(1):51–56. doi:10.1159/000069504

Gonzalez-Alegre P, Kelkar P, Rodnitzky RL. Isolated high-frequency jaw tremor relieved by botulinum toxin injections. *Mov Disord.* 2006;21(7):1049–1050. doi:10.1002/mds.20878

Goodman RR, Kim B, McClelland S 3rd, Senatus PB, Winfield LM, Pullman SL, Yu Q , Ford B, McKhann GM 2nd. Operative techniques and morbidity with subthalamic nucleus deep brain stimulation in 100 consecutive patients with advanced Parkinson's disease. *J Neurol Neurosurg Psychiatry.* 2006;77(1):12–17. doi:10.1136/jnnp.2005.069161

Goto S, Nishikawa S, Mita S, Ushio Y. Impact of posterior GPI pallidotomy on leg tremor in Parkinson's disease. *Stereotact Funct Neurosurg.* 2002;78(2):64–69. doi:10.1159/000068013

Gunal DI, Af ar N, Bekiroglu N, Aktan S. New alternative agents in essential tremor therapy: double-blind placebo-controlled study of alprazolam and acetazolamide. *Neurol Sci.* 2000;21(5):315–317. doi:10.1007/s100720070069

Habib-ur-Rehman. Diagnosis and management of tremor. *Arch Intern Med.* 2000;160(16):2438–2444. doi:10.1001/archinte.160.16.2438

Hallett M, Ravits J, Dubinsky RM, Gillespie MM, Moinfar A. A double-blind trial of isoniazid for essential tremor and other action tremors. *Mov Disord.* 1991;6(3):253–256. doi:10.1002/mds.870060310

Hamel W, Herzog J, Kopper F, Pinsker M, Weinert D, Müller D, Krack P, Deuschl G, Mehdorn HM. Deep brain stimulation in the subthalamic area is more effective than nucleus ventralis intermedius stimulation for bilateral intention tremor. *Acta Neurochir (Wien).* 2007;149(8):749–758; discussion 758. doi:10.1002/mds.870060310

Hammond ER, Kerr DA. Ethanol responsive tremor in a patient with multiple sclerosis. *Arch Neurol.* 2008;65(1):142–143. doi:10.1001/archneurol.2007.13

Handforth A, Martin FC. Pilot efficacy and tolerability: a randomized, placebo-controlled trial of levetiracetam for essential tremor. *Mov Disord.* 2004;19(10):1215–1221. doi:10.1002/mds.20147

Henderson JM, Portmann L, Van Melle G, Haller E, Ghika JA. Propranolol as an adjunct therapy for hyperthyroid tremor. *Eur Neurol.* 1997;37(3):182–185. doi:10.1159/000117431

Higuchi Y, Iacono RP. Surgical complications in patients with Parkinson's disease after posteroventral pallidotomy. *Neurosurgery.* 2003;52(3):558–571. doi:10.1227/01.NEU.0000047817.60776.5C

Hooper AK, Okun MS, Foote KD, Fernandez HH, Jacobson C, Zeilman P, Romrell J, Rodriguez RL. Clinical cases where lesion therapy was chosen over deep brain stimulation. *Stereotact Funct Neurosurg.* 2008;86(3):147–152 doi:10.1159/000120426

Houdayer E, Devanne H, Tyvaert L, Defebvre L, Derambure P, Cassim F. Low frequency repetitive transcranial magnetic stimulation over premotor cortex can improve cortical tremor. *Clin Neurophysiol.* 2007;118(7):1557–1562. doi:10.1016/j.clinph.2007.04.014

Jiménez-Jiménez FJ, Garcia-Ruiz PJ, Cabrera-Valdivia F. Nicardipine versus propranolol in essential tumor. *Acta Neurol (Napoli).* 1994;16(4):184–188.

Katayama Y, Kasai M, Oshima H, Fukaya C, Yamamoto T, Ogawa K, Mizutani T. Subthalamic nucleus stimulation for Parkinson disease: benefits observed in levodopa-intolerant patients. *J Neurosurg.* 2001;95(2):213–221.

Kim MC, Son BC, Miyagi Y, Kang JK. Vim thalamotomy for Holmes' tremor secondary to midbrain tumour. *J Neurol Neurosurg Psychiatry.* 2002;73(4):453–455. doi:10.1136/jnnp.73.4.453

Kimber TE, Brophy BP, Thompson PD. Ataxic arm movements after thalamotomy for Parkinsonian tremor. *J Neurol Neurosurg Psychiatry.* 2003;74(2):258–259. doi:10.1136/jnnp.74.2.258

Klebe S, Stolze H, Grensing K, Volkmann J, Wenzelburger R, Deuschl G. Influence of alcohol on gait in patients with essential tremor. *Neurology.* 2005;65(1):96–101. doi:10.1212/01.wnl.0000167550.97413.1f

Koller WC. Pharmacologic treatment of parkinsonian tremor. *Arch Neurol.* 1986;43(2):126–127.

Koller WC, Herbster G. Adjuvant therapy of parkinsonian tremor. *Arch Neurol.* 1987;44(9):921–923.

Kondziolka D, Ong JG, Lee JY, Moore RY, Flickinger JC, Lunsford LD. Gamma Knife thalamotomy for essential tremor. *J Neurosurg.* 2008;108(1):111–117. doi:10.3171/JNS/2008/108/01/0111

Krack P, Benazzouz A, Pollak P, Limousin P, Piallat B, Hoffmann D, Xie J, Benabid AL. Treatment of tremor in Parkinson's disease by subthalamic nucleus stimulation. *Mov Disord.* 1998;13(6):907–914. doi:10.1002/mds.870130608

Krauss JK, Mohadjer M, Nobbe F, Mundinger F. The treatment of posttraumatic tremor by stereotactic surgery. Symptomatic and functional outcome in a series of 35 patients. *J Neurosurg.* 1994;80(5):810–819.

Kuehler A, Henrich G, Schroeder U, Conrad B, Herschbach P, Ceballos-Baumann A. A novel quality of life instrument for deep brain stimulation in movement disorders. *J Neurol Neurosurg Psychiatry.* 2003;74(8):1023–1030. doi:10.1136/jnnp.74.8.1023

Lee JY, Kondziolka D. Thalamic deep brain stimulation for management of essential tremor. *J Neurosurg.* 2005;103(3):400–403.

Lee KS, Kim JS, Kim JW, Lee WY, Jeon BS, Kim D. A multicenter randomized crossover multiple-dose comparison study of arotinolol and propranolol in essential tremor. *Parkinson Relat Disord.* 2003;9(6):341–347. doi:10.1016/S1353-8020(03)00029-4

Leuthardt EC, Schalk G, Wolpaw JR, Ojemann JG, Moran DW. A brain–computer interface using electrocorticographic signals in humans. *J Neural Eng.* 2004;1(2):63–71. doi:10.1088/1741-2560/1/2/001

Liou LM, Shih PY. Successful treatment of rubral tremor by high-dose trihexyphenidyl: a case report. *Kaohsiung J Med Sci.* 2006;22(3):149–153.

Llumiguano C, Dóczi T, Baths I. Microeletrode guided stereotactic pallidotomy and pallido-thalamotomy for treatment of Parkinson's disease. *Neurocirugia (Astur).* 2006;17(5):420–432.

Lyons KE, Pahwa R, Comella CL, Eisa MS, Elble RJ, Fahn S, Jankovic J, Juncos JL, Koller WC, Ondo WG, Sethi KD, Stern MB, Tanner CM, Tintner R, Watts RL. Benefits and risks of pharmacological treatments for essential tremor. *Drug Saf.* 2003;26(7):461–481. doi:10.2165/00002018-200326070-00003

Lyons KE, Wilkinson SB, Overman J, Pahwa R. Surgical and hardware complications of subthalamic stimulation: a series of 160 procedures. *Neurology.* 2004;63(4):612–616.

Maeda F, Keenan JP, Tormos JM, Topka H, Pascual-Leone A. Modulation of corticospinal excitability by repetitive transcranial magnetic stimulation. *Clin Neurophysiol.* 2000;111(5):800–805. doi:10.1016/S1388-2457(99)00323-5

Manto M, Rocon E, Pons J, Belda JM, Camut S. Evaluation of a wearable orthosis and an associated algorithm for tremor suppression. *Physiol Meas.* 2007;28(4):415–425. doi:10.1088/0967-3334/28/4/007

Mathieu D, Kondziolka D, Niranjan A, Flickinger J, Lunsford LD. Gamma knife thalamotomy for multiple sclerosis tremor. *Surg Neurol.* 2007;68(4):394–399. doi:10.1016/j.surneu.2006.11.049

Metzer WS, Paige SR, Newton JE. Inefficacy of propranolol in attenuation of drug-induced parkinsonian tremor. *Mov Disord.* 1993;8(1):43–46. doi:10.1002/mds.870080108

Muenter MD, Daube JR, Caviness JN, Miller PM. Treatment of essential tremor with methazolamide. *Mayo Clin Proc.* 1991;66(10):991–997.

Niranjan A, Jawahar A, Kondziolka D, Lunsford LD. A comparison of surgical approaches for the management of tremor: radiofrequency thalamotomy, gamma knife thalamotomy and thalamic stimulation. *Stereotact Funct Neurosurg.* 1999;72(2–4):178–184. doi:10.1159/000029723

Ondo WG, Jimenez JE, Vuong KD, Jankovic J. An open-label pilot study of levetiracetam for essential tremor. *Clin Neuropharmacol.* 2004;27(6):274–277. doi:10.1097/00002826-200411000-00004

Ondo WG, Jankovic J, Connor GS, Pahwa R, Elble R, Stacy MA, Koller WC, Schwarzman L, Wu SC, Hulihan JF; Topiramate Essential Tremor Study Investigators. Topiramate in essential tremor: a double-blind, placebo-controlled trial. *Neurology.* 2006;66(5):672–677.

Pahwa R, Lyons KE. Essential tremor: differential diagnosis and current therapy. *Am J Med.* 2003;115(2):134–142. doi:10.1016/S0002-9343(03)00259-6

Papavassiliou E, Rau G, Heath S, Abosch A, Barbaro NM, Larson PS, Lamborn K, Starr PA. Thalamic deep brain stimulation for essential tremor: relation of lead location to outcome. *Neurosurgery.* 2004;54(5):1120–1129. doi:10.1227/01.NEU.0000119329.66931.9E

Patel NK, Heywood P, O'Sullivan K, McCarter R, Love S, Gill SS. Unilateral subthalamotomy in the treatment of Parkinson's disease. *Brain.* 2003;126(Pt 5):1136–1145. doi:10.1093/brain/awg111

Plaha P, Ben-Shlomo Y, Patel NK, Gill SS. Stimulation of the caudal zona incerta is superior to stimulation of the subthalamic nucleus in improving contralateral parkinsonism. *Brain.* 2006 Jul;129(Pt 7):1732–1747. doi:10.1093/brain/awl127

Raj V, Landess JS, Martin PR. Oxcarbazepine use in essential tremor. *Ann Pharmacother.* 2006;40(10):1876–1879. doi:10.1345/aph.1H131

Rice GP, Lesaux J, Vandervoort P, Macewan L, Ebers GC. Ondansetron, a 5-HT3 antagonist, improves cerebellar tremor. *J Neurol Neurosurg Psychiatry.* 1997;62(3):282–284.

Rocon E, Belda-Lois JM, Ruiz AF, Manto M, Moreno JC, Pons JL. Design and validation of a rehabilitation robotic exoskeleton for tremor assessment and suppression. *IEEE Trans Neural Syst Rehabil Eng.* 2007;15(3):367–378.

Rocon E, Manto M, Pons J, Camut S, Belda JM. Mechanical suppression of essential tremor. *Cerebellum.* 2007;6(1):73–78. doi:10.1080/14734220601103037

Rodrigues JP, Edwards DJ, Walters SE, Byrnes ML, Thickbroom GW, Stell R, Mastaglia FL. Blinded placebo crossover study of gabapentin in primary orthostatic tremor. *Mov Disord.* 2006;21(7):900–905. doi:10.1002/mds.20830

Sansur CA, Frysinger RC, Pouratian N, Fu KM, Bittl M, Oskouian RJ, Laws ER, Elias WJ. Incidence of symptomatic hemorrhage after stereotactic electrode placement. *J Neurosurg.* 2007;107(5):998–1003. doi:10.3171/jns.2007.107.5.998

Sasso E, Perucca E, Fava R, Calzetti S. Primidone in the long-term treatment of essential tremor: a prospective study with computerized quantitative analysis. *Clin Neuropharmacol.* 1990;13(1):67–76.

Schuurman PR, Bruins J, Merkus MP, Bosch DA, Speelman JD. A comparison of neuropsychological effects of thalamotomy and thalamic stimulation. *Neurology.* 2002;59(8):1232–1239.

Seijo FJ, Alvarez-Vega MA, Gutierrez JC, Fdez-Glez F, Lozano B. Complications in subthalamic nucleus stimulation surgery for treatment of Parkinson's disease. Review of 272 procedures. *Acta Neurochir (Wien).* 2007;149(9):867–875. doi:10.1007/s00701-007-1267-1

Serrano-Dueñas M. Use of primidone in low doses (250 mg/day) versus high doses (750 mg/day) in the management of essential tremor. Double-blind comparative study with one-year follow-up. *Parkinsonism Relat Disord.* 2003;10(1):29–33. doi:10.1016/S1353-8020(03)00070-1

Siderowf A, Gollump SM, Stern MB, Baltuch GH, Riina HA. Emergence of complex, involuntary movements after gamma knife radiosurgery for essential tremor. *Mov Disord.* 2001;16(5):965–967. doi:10.1002/mds.1178

Soulas T, Gurruchaga JM, Palfi S, Cesaro P, Nguyen JP, Fénelon G. Attempted and completed suicides after subthalamic nucleus stimulation for Parkinson's disease. *J Neurol Neurosurg Psychiatry.* 2008. doi:10.1136/jnnp.2007.130583

Striano P, Coppola A, Vacca G, Zara F, Brescia Morra V, Orefice G, Striano S. Levetiracetam for cerebellar tremor in multiple sclerosis: an open-label pilot tolerability and efficacy study. *J Neurol.* 2006;253(6):762–766.

Teodorescu HL, Kandel A, Hall LO. Report of research activities in fuzzy AI and medicine at USF CSE. *Artif Intell Med.* 2001;21(1–3):177–183. doi:10.1016/S0933-3657(00)00083-X

Velasco F, Jiménez F, Pérez ML, Carrillo-Ruiz JD, Velasco AL, Ceballos J, Velasco M. Electrical stimulation of the prelemniscal radiation in the treatment of Parkinson's disease: an old target revised with new techniques. *Neurosurgery.* 2001;49(2):293–306. doi:10.1097/00006123-200108000-00009

Vergani F, Landi A, Antonini A, Sganzerla EP. Bilateral subthalamic deep brain stimulation in a patient with Parkinson's disease who had previously undergone thalamotomy and autologous adrenal grafting in the caudate nucleus: case report. *Neurosurgery.* 2006;59(5):E1140; discussion E1140. doi:10.1227/01.NEU.0000245585.93284.15

Vidailhet M, Dupel C, Lehéricy S, Remy P, Dormont D, Serdaru M, Jedynak P, Veber H, Samson Y, Marsault C, Agid Y. Dopaminergic dysfunction in midbrain dystonia: anatomo-clinical study using 3-dimensional magnetic resonance imaging and fluorodopa F 18 positron emission tomography. *Arch Neurol.* 1999;56(8):982–989. doi:10.1001/archneur.56.8.982

Wills AJ, Brusa L, Wang HC, Brown P, Marsden CD. Levodopa may improve orthostatic tremor: case report and trial of treatment. *J Neurol Neurosurg Psychiatry.* 1999;66(5):681–684.

Yap L, Kouyialis A, Varma TR. Stereotactic neurosurgery for disabling tremor in multiple sclerosis: thalamotomy or deep brain stimulation? *Br J Neurosurg.* 2007;21(4):349–354. doi:10.1080/02688690701544002

Yetimalar Y, Irtman G, Kurt T, Basoğlu M. Olanzapine versus propranolol in essential tremor. *Clin Neurol Neurosurg.* 2005;108(1):32–35. doi:10.1016/j.clineuro.2005.01.002

Young RF, Jacques S, Mark R, Kopyov O, Copcutt B, Posewitz A, Li F. Gamma knife thalamotomy for treatment of tremor: long-term results. *J Neurosurg.* 2000;93(Suppl 3):128–135.

Zesiewicz TA, Elble R, Louis ED, Hauser RA, Sullivan KL, Dewey RB Jr, Ondo WG, Gronseth GS, Weiner WJ; Quality Standards Subcommittee of the American Academy of Neurology. Practice parameter: therapies for essential tremor: report of the Quality Standards Subcommittee of the American Academy of Neurology. *Neurology.* 2005;64(12):2008–2020. doi:10.1212/01.WNL.0000163769.28552.CD

Zesiewicz TA, Ward CL, Hauser RA, Salemi JL, Siraj S, Wilson MC, Sullivan KL. A pilot, double-blind, placebo-controlled trial of pregabalin (Lyrica) in the treatment of essential tremor. *Mov Disord.* 2007;22(11):1660–1663.

Zeuner KE, Molloy FM, Shoge RO, Goldstein SR, Wesley R, Hallett M. Effect of ethanol on the central oscillator in essential tremor. *Mov Disord.* 2003;18(11):1280–1285. doi:10.1002/mds.10553

• • • •

Annex 1

ANNEX 1

Extended ADL Scale according to Nourie and Lincoln (Nottingham Stroke Score)				
SCORING: **0= NO/WITH HELP** **1= ON MY OWN WITH DIFFICULTY/** **ON MY OWN**	**NO**	**WITH HELP**	**ON MY OWN WITH DIFFICULTY**	**ON MY OWN**
Mobility				
–do you walk around outside?				
–do you climb stairs?				
–do you get in and out of the car?				
–do you walk over uneven ground?				
–do you cross roads?				
–do you travel on public transport?				
In the kitchen				
–do you manage to feed yourself?				
–do you manage to make yourself a hot drink?				
–do you take hot drinks from one room to another?				
–do you do the washing up?				

–do you make yourself a hot snack?				
Domestics tasks				
–do you manage your own money when you are out?				
–do you wash small items of clothing?				
–do you do your own housework?				
–do you do your own shopping?				
–do you do a full clothes wash?				
Leisure activities				
–do you read newspapers or books?				
–do you use the telephone?				
–do you write letters?				
–do you go out socially?				
–do you manage your own garden?				
–do you drive a car?				
Total score =				

From: Nouri and Lincoln (1987).

Schwab and England Activities of Daily Living scale

100% = Completely independent. Able to do all chores without slowness, difficulty or impairment. Essentially normal. Unaware of any difficulty.

90% = Completely independent. Able to do all chores with some degree of slowness, difficulty and impairment. Might take twice as long. Beginning to be aware of difficulty.

80% = Completely independent in most chores. Takes twice as long. Conscious of difficulty and slowness.

70% = Not completely independent. More difficulty with some chores. Three to four times as long in some. Must spend a large part of the day with chores.

60% = Some dependency. Can do most chores, but exceedingly slowly and with much effort. Errors; some impossible.

50% = More dependent. Help with half, slower, etc. Difficulty with everything.

40% = Very dependent. Can assist with all chores, but few alone.

30% = With effort, now and then does a few chores alone or begins alone. Much help needed.

20% = Nothing alone. Can be a slight help with some chores. Severe invalid.

10% = Totally dependent, helpless. Complete invalid.

0% = Vegetative functions such as swallowing, bladder, and bowel functions are not functioning. Bedridden.

From: Schwab and England Activities of Daily Living (Gillingham and Donaldson, 1969).

Annex 2

ANNEX 2

Clinical Tremor Rating Scale
1–10 Tremor Rate tremor at rest in items 1 and 2. Rate tremor with posture holding in items 3–10. Abbreviations: UE: upper extremities, LE: lower extremities. score: 0 = none; 1 = slight (amplitude < 0.5 cm), may be intermittent; 2 = moderate amplitude (0.5–1 cm), may be intermittent; 3 = marked amplitude (1–2 cm); 4 = severe amplitude (>2 cm).
1 Head
2 Trunk
3 UE: arms outstretched, wrist mildly extended, fingers spread apart
4 LE: legs flexed at hips and knees
5 Foot dorsiflexed
6 Tongue: when protruded
7 Head and trunk: when sitting or standing
8 Rate tremor with action and intention
9 UE: finger to nose and other actions
10 LE: toe to finger in flexed posture

11 Handwriting

Have patient write the standard sentence: "This is a sample of my best handwriting", sign his or her name and write the date.

score: 0 normal; 1= mildly abnormal, slight untidy, tremulous; 2= moderate abnormal, legible but with considerable tremor; 3 marked abnormal, illegible; 4 severely abnormal, unable to keep pencil or pen on paper without holding hand down with the other hand.

12–14 Drawings

Ask the patient to join both points of the various drawings without crossing the lines. Test each hand, beginning with the lesser involved, without leaning the hand or arm on the table

score: 0 = normal; 1 = slightly tremulous, may cross lines occasionally; 2 = moderately tremulous or crosses lines frequently; 3 = accomplishes the task with great difficulty, many errors; 4 = unable to complete drawing.

15 Pouring

Use firm plastic cups (8 cm tall), filled with water to 1 cm fro top. Ask the patient to pour water from one cup to another. Test each hand separately.

16 Speaking

This includes spastic dysphonia if present

score: 0 = normal; 1 = mild voice tremulousness when "nervous" only; 2 = mild voice tremulousness, constant; 3 = moderate voice tremor; 4 = severe voice tremor, some word difficult to understand.

17 Feeding

Other than liquids

score: 0 = normal; 1 = mildly abnormal, can bring all solid to mouth, spilling only rarely; 2 = moderately abnormal, frequent spills of peas and similar foods, may bring head at least half way to meet food; 3 = markedly abnormal, unable to cut or uses two hands to feed; 4 = severely abnormal, needs help to feed.

18 Bringing liquids to mouth

score: 0 = normal; 1 = mildly abnormal, can still use a spoon, is completely full; 2 = moderately abnormal, unable to use a spoon, uses cups or glasses; 3 = markedly abnormal, can drink from cup or glass, but needs help two hands; 4 = severely abnormal, must use a straw.

19 Hygiene

  score: 0 = normal; 1 = mildly abnormal, able to do everything but is more careful than the average person; 2 = moderately abnormal, able to do everything but with errors, uses electric razor because of tremor; 3 = markedly abnormal, unable to do most fine tasks, such as putting on lipstick or shaving (even with electric shaver), unless using two hands; 4 = severely abnormal, unable to do any fine movement tasks.

20 Dressing

  score: 0 = normal; 1 = mildly abnormal, able to do everything but is more careful than the average person; 2 = moderately abnormal, able to do everything but with errors; 3 = markedly abnormal, needs some assistance with buttoning or other activities, such as tying shoelaces; 4 = severely abnormal, requires assistance even for gross motor activities.

21 Writing

  score: 0 = normal; 1 = mildly abnormal, legible, continue to write letters; 2 = moderately abnormal, legible but no longer writes letters; 3 = markedly abnormal, illegible; 4 = severely abnormal, unable to sign checks or other documents requiring signature

22 Working

  score: 0 = tremor does not interfere with job; 1 = able to work, but needs to be more careful than the average person; 2 = able to do everything but with errors, poorer than usual performance because of tremor; 3 = unable to do regular job, may have changed to a different job because of tremor, tremor limits housework, such as ironing; 4 = unable to do any outside job, housework very limited.

Total score

From: Fahn et al. (1988).

Annex 3

ANNEX 3
Unified Parkinson's Disease Rating Scale (UDPRS)

I. MENTATION, BEHAVIOR, AND MOOD

1. Intellectual impairment

0 = None.

1 = Mild. Consistent forgetfulness with partial recollection of events and no other difficulties.

2 = Moderate memory loss, with disorientation and moderate difficulty handling complex problems. Mild but definite impairment of function at home with need of occasional prompting.

3 = Severe memory loss with disorientation for time and often to place. Severe impairment in handling problems.

4 = Severe memory loss with orientation preserved to person only. Unable to make judgements or solve problems. Requires much help with personal care. Cannot be left alone at all.

2. Thought disorder (due to dementia or drug intoxication)

0 = None.

1 = Vivid dreaming.

2 = "Benign" hallucinations with insight retained.

3 = Occasional to frequent hallucinations or delusions; without insight; could interfere with daily activities.

4 = Persistent hallucinations, delusions, or florrid psychosis. Not able to care for self.

3. Depression

1 = Periods of sadness or guilt greater than normal, never sustained for days or weeks.

2 = Sustained depression (1 week or more).

3 = Sustained depression with vegetative symptoms (insomnia, anorexia, weight loss, loss of interest).

4 = Sustained depression with vegetative symptoms and suicidal thoughts or intent.

4. Motivation/initiative

0 = Normal.

1 = Less assertive than usual; more passive.

2 = Loss of initiative or disinterest in elective (nonroutine) activities.

3 = Loss of initiative or disinterest in day to day (routine) activities.

4 = Withdrawn, complete loss of motivation.

II. ACTIVITIES OF DAILY LIVING (for both "on" and "off")

5. Speech

0 = Normal.

1 = Mildly affected. No difficulty being understood.

2 = Moderately affected. Sometimes asked to repeat statements.

3 = Severely affected. Frequently asked to repeat statements.

4 = Unintelligible most of the time.

6. Salivation

0 = Normal.

1 = Slight but definite excess of saliva in mouth; may have nighttime drooling.

2 = Moderately excessive saliva; may have minimal drooling.

3 = Marked excess of saliva with some drooling.

4 = Marked drooling, requires constant tissue or handkerchief.

7. Swallowing

0 = Normal.

1 = Rare choking.

2 = Occasional choking.

3 = Requires soft food.

4 = Requires NG tube or gastrotomy feeding.

8. Handwriting

0 = Normal.

1 = Slightly slow or small.

2 = Moderately slow or small; all words are legible.

3 = Severely affected; not all words are legible.

4 = The majority of words are not legible.

9. Cutting food and handling utensils

0 = Normal.

1 = Somewhat slow and clumsy, but no help needed.

2 = Can cut most foods, although clumsy and slow; some help needed.

3 = Food must be cut by someone, but can still feed slowly.

4 = Needs to be fed.

10. Dressing

0 = Normal.

1 = Somewhat slow, but no help needed.

2 = Occasional assistance with buttoning, getting arms in sleeves.

3 = Considerable help required, but can do some things alone.

4 = Helpless.

11. Hygiene

0 = Normal.

1 = Somewhat slow, but no help needed.

2 = Needs help to shower or bathe; or very slow in hygienic care.

3 = Requires assistance for washing, brushing teeth, combing hair, going to bathroom.

4 = Foley catheter or other mechanical aids.

12. Turning in bed and adjusting bed clothes

0 = Normal.

1 = Somewhat slow and clumsy, but no help needed.

2 = Can turn alone or adjust sheets, but with great difficulty.

3 = Can initiate, but not turn or adjust sheets alone.

4 = Helpless.

13. Falling (unrelated to freezing)

0 = None.

1 = Rare falling.

2 = Occasionally falls, less than once per day.

3 = Falls an average of once daily.

4 = Falls more than once daily.

14. Freezing when walking

0 = None.

1 = Rare freezing when walking; may have starthesitation.

2 = Occasional freezing when walking.

3 = Frequent freezing. Occasionally falls from freezing.

4 = Frequent falls from freezing.

15. Walking

0 = Normal.

1 = Mild difficulty. May not swing arms or may tend to drag leg.

2 = Moderate difficulty, but requires little or no assistance.

3 = Severe disturbance of walking, requiring assistance.

4 = Cannot walk at all, even with assistance.

16. Tremor (symptomatic complaint of tremor in any part of body)

0 = Absent.

1 = Slight and infrequently present.

2 = Moderate; bothersome to patient.

3 = Severe; interferes with many activities.

4 = Marked; interferes with most activities.

17. Sensory complaints related to parkinsonism

0 = None.

1 = Occasionally has numbness, tingling, or mild aching.

2 = Frequently has numbness, tingling, or aching; not distressing.

3 = Frequent painful sensations.

4 = Excruciating pain.

III. MOTOR EXAMINATION

18. Speech

0 = Normal.

1 = Slight loss of expression, diction and/or volume.

2 = Monotone, slurred but understandable; moderately impaired.

3 = Marked impairment, difficult to understand.

4 = Unintelligible.

19. Facial expression

0 = Normal.

1 = Minimal hypomimia, could be normal "Poker Face."

2 = Slight but definitely abnormal diminution of facial expression

3 = Moderate hypomimia; lips parted some of the time.

4 = Masked or fixed facies with severe or complete loss of facial expression; lips parted 1/4 inch or more.

20. Tremor at rest (head, upper and lower extremities)

0 = Absent.

1 = Slight and infrequently present.

2 = Mild in amplitude and persistent. Or moderate in amplitude, but only intermittently present.

3 = Moderate in amplitude and present most of the time.

4 = Marked in amplitude and present most of the time.

21. Action or postural tremor of hands

0 = Absent.

1 = Slight; present with action.

2 = Moderate in amplitude, present with action.

3 = Moderate in amplitude with posture holding as well as action.

4 = Marked in amplitude; interferes with feeding.

22. Rigidity (judged on passive movement of major joints with patient relaxed in sitting position; cog-wheeling to be ignored.)

0 = Absent.

1 = Slight or detectable only when activated by mirror or other movements.

2 = Mild to moderate.

3 = Marked, but full range of motion easily achieved.

4 = Severe, range of motion achieved with difficulty.

23. Finger taps (patient taps thumb with index finger in rapid succession)

0 = Normal.

1 = Mild slowing and/or reduction in amplitude.

2 = Moderately impaired. Definite and early fatiguing. May have occasional arrests in movement.

3 = Severely impaired. Frequent hesitation in initiating movements or arrests in ongoing movement.

4 = Can barely perform the task.

24. Hand movements (patient opens and closes hands in rapid succession)

0 = Normal.

1 = Mild slowing and/or reduction in amplitude.

2 = Moderately impaired. Definite and early fatiguing. May have occasional arrests in movement.

3 = Severely impaired. Frequent hesitation in initiating movements or arrests in ongoing movement.

4 = Can barely perform the task.

25. Rapid alternating movements of hands (pronation–supination movements of hands, vertically, and horizontally, with as large an amplitude as possible, both hands simultaneously)

0 = Normal.

1 = Mild slowing and/or reduction in amplitude.

2 = Moderately impaired. Definite and early fatiguing. May have occasional arrests in movement.

3 = Severely impaired. Frequent hesitation in initiating movements or arrests in ongoing movement.

4 = Can barely perform the task.

26. Leg agility (Patient taps heel on the ground in rapid succession picking up entire leg. Amplitude should be at least 3 in.)

0 = Normal.

1 = Mild slowing and/or reduction in amplitude.

2 = Moderately impaired. Definite and early fatiguing. May have occasional arrests in movement.

3 = Severely impaired. Frequent hesitation in initiating movements or arrests in ongoing movement.

4 = Can barely perform the task.

27. Arising from chair (patient attempts to rise from a straightbacked chair, with arms folded across chest)

0 = Normal.

1 = Slow; or may need more than one attempt.

2 = Pushes self up from arms of seat.

3 = Tends to fall back and may have to try more than one time, but can get up without help.

4 = Unable to arise without help.

28. *Posture*

0 = Normal erect.

1 = Not quite erect, slightly stooped posture; could be normal for older person.

2 = Moderately stooped posture, definitely abnormal; can be slightly leaning to one side.

3 = Severely stooped posture with kyphosis; can be moderately leaning to one side.

4 = Marked flexion with extreme abnormality of posture.

29. *Gait*

0 = Normal.

1 = Walks slowly, may shuffle with short steps, but no festination (hastening steps) or propulsion.

2 = Walks with difficulty, but requires little or no assistance; may have some festination, short steps, or propulsion.

3 = Severe disturbance of gait, requiring assistance.

4 = Cannot walk at all, even with assistance.

30. *Postural stability (response to sudden, strong posterior displacement produced by pull on shoulders while patient erect with eyes open and feet slightly apart; patient is prepared)*

0 = Normal.

1 = Retropulsion, but recovers unaided.

2 = Absence of postural response; would fall if not caught by examiner.

3 = Very unstable, tends to lose balance spontaneously.

4 = Unable to stand without assistance.

31. *Body bradykinesia and hypokinesia (combining slowness, hesitancy, decreased arm swing, small amplitude, and poverty of movement in general)*

0 = None.

1 = Minimal slowness, giving movement a deliberate character; could be normal for some persons. Possibly reduced amplitude.

2 = Mild degree of slowness and poverty of movement which is definitely abnormal. Alternatively, some reduced amplitude.

3 = Moderate slowness, poverty or small amplitude of movement.

4 = Marked slowness, poverty or small amplitude of movement.

IV. COMPLICATIONS OF THERAPY (in the past week)

A. DYSKINESIAS

32. Duration: What proportion of the waking day are dyskinesias present? (Historical information)
0 = None
1 = 1–25% of day.
2 = 26–50% of day.
3 = 51–75% of day.
4 = 76–100% of day.

33. Disability: How disabling are the dyskinesias? (Historical information; may be modified by office examination)
0 = Not disabling.
1 = Mildly disabling.
2 = Moderately disabling.
3 = Severely disabling.
4 = Completely disabled.

34. Painful dyskinesias: How painful are the dyskinesias?
0 = No painful dyskinesias.
1 = Slight.
2 = Moderate.
3 = Severe.
4 = Marked.

35. Presence of early morning dystonia (historical information)
0 = No
1 = Yes

B. CLINICAL FLUCTUATIONS

36. Are "off" periods predictable?
0 = No
1 = Yes

37. Are "off" periods unpredictable?
0 = No
1 = Yes

38. Do "off" periods come on suddenly, within a few seconds?
0 = No
1 = Yes

39. What proportion of the waking day is the patient "off" on average?
0 = None
1 = 1–25% of day.
2 = 26–50% of day.
3 = 51–75% of day.
4 = 76–100% of day.

C. OTHER COMPLICATIONS

40. Does the patient have anorexia, nausea, or vomiting?
0 = No
1 = Yes

41. Any sleep disturbances, such as insomnia or hypersomnolence?
0 = No
1 = Yes

42. Does the patient have symptomatic orthostasis?
(Record the patient's blood pressure, height, and weight on the scoring form)
0 = No
1 = Yes

V. MODIFIED HOEHN AND YAHR STAGING

STAGE 0 = No signs of disease.
STAGE 1 = Unilateral disease.
STAGE 1.5 = Unilateral plus axial involvement.

STAGE 2 = Bilateral disease, without impairment of balance.

STAGE 2.5 = Mild bilateral disease, with recovery on pull test.

STAGE 3 = Mild to moderate bilateral disease; some postural instability; physically independent.

STAGE 4 = Severe disability; still able to walk or stand unassisted.

STAGE 5 = Wheelchair bound or bedridden unless aided.

From: Fahn et al. (1987).

List of Abbreviations

5-FU: 5-fluorouracil

5HT: 5-hydroxytryptamine

6-OHDA: 6-hydroxydopamine

9HPT: Nine Hole Peg Test

A: adenosine

Ach: acetylcholine

ADL: Activity of Daily Living

Ago: agonist

AIN: anterior interpositus nucleus

AMPA: α-amino-3-hydroxy-5-methylisoazol-4-propionate

Antago: antagonist

ATP: adenosine triphosphate

ATPase: adenosine triphosphatase

BCA: beta-carboline alkaloids

BCIs: brain–computer interfaces

BG: basal ganglia

BHFSD: bilateral high-frequency synchronous discharges

BSN: body sensor networks

C7–C8: cervical roots 7–8

Ca: calcium

CBD: corticobasal ganglionic degeneration

CC: cerebellar cortex

CD: cervical dystonia

cf: climbing fibers

CI: confidence interval

CIDP: chronic inflammatory demyelinating polyneuropathy

Cl: chloride

CM/Pf: centromedian/parafascicular nucleus of the thalamus

CM: centromedian nucleus

CN: cerebellar nuclei

CNS: central nervous system

CO: carbon monoxide

CRPS: complex regional pain syndrome

CRS: clincal rating scale

CS: cross-spectrum

CSF: cerebrospinal fluid

CT: computerized tomography

CWT: continuous wavelet transform

DA: dopamine

DAT: dopamine transporter

DBS: deep brain stimulation

DR: dorsal root ganglia

DSCT: dorsal spinocerebellar tract

DTI: diffusion tensor MRI

DWT: discrete wavelet transform

E: electrical potential across the membrane

ECoG: electrocorticography

ECR: extenson carpi radialis

EDC: extensor digitorum communis

EEG: electroencephalography

EMD: empirical mode decomposition

EMG: electromyography

EPSP: excitatory postsynaptic potentials

EPT: enhanced physiological tremor

ET: essential tremor

ETV: essential tremor of voice

F: Faraday unit

FCMTE: familial cortical myoclonic tremor with epilepsy

FCR: flexor carpi radialis

FFT: fast Fourier transform

FMR1: fragile X mental retardation 1 gene

fMRI: functional magnetic resonance imaging

FT: Fourier transform_

FXTAS: fragile-X-associated tremor/ataxia syndrome

GABA: amino-butyric acid

GC: granule cells

Gly: glycine

GP: globus pallidus

GPe: globus pallidus—external segment

GPi: globus pallidus—internal segment

GTP: guanosine triphosphate

HIV: human immunodeficiency virus

HS: Hilbert spectrum

Hz: Hertz

Ia: Ia sensory afferent fibers

IC: internal capsula

ICH: intracranial hemorrhage

III: third ventricle

IMFs: intrinsic mode functions

IO: inferior olive

ION: inferior olivary nucleus

IPSP: inhibitory postsynaptic potential

K: potassium

LB: Lewy body

LBD: Lewy body disease

LLR: long-latency reflexes

LMN: lower motoneuron

LN: Lewy neuritis

LTP: long-term potentiation

M1: primary motor cortex

MEA: midbrain extrapyramidal area

MEG: magnetoencephalography

mf: mossy fibers

MGUS: monoclonal gammopathy of undetermined significance

MHS: marginal Hilbert spectrum

mM: millimole

MN α: alpha motoneuron

MN γ: gamma motoneuron

MPTP: 1-methyl-4-phenyl-1,2,3,6-tetrahydropyridine

MRI: magnetic resonance imaging

ms: millisecond

MS: multiple sclerosis

MSA: multiple system atrophy

MU: motor unit

MUAP: motor unit action potentials

mV: millivolt

MVC: maximal voluntary contraction

Na: sodium

NAA/tCr: cerebellar *N*-acetylaspartate/total creatine

NIP: neural input processor

nm: nanometer

NMDA: *N*-methyl-D-aspartate

NRTP: nucleus reticularis tegmenti pontis

OPCA: olivopontocerebellar atrophy

OT: orthostatic tremor

PD: Parkinson's disease

PE: parallel viscous element

PET: positron emission tomography

PF: parallel fibers

PMC: premotor cortex

PN: peripheral neuropathy

PN: Purkinje neurons

POT: primary orthostatic tremor

PPC: posterior parietal cortex

PSD: power spectral density

PSP: progressive supranuclear palsy

PT: physiological tremor

Put: putamen

PVP: posteroventral pallidotomy

PWT: primary writing tremor

R: resistance

RAPRL: prelemniscal radiation

RAS: reticular activating system

RC: Renshaw cells

REM: rapid eye movements

RetST: reticulospinal tract

RFL: radiofrequency lesioning

RN: red nucleus

rTMS: repetitive transcranial magnetic stimulation

RuST: rubrospinal tract

SE: series elastic component

SEMG: surface EMG electrodes

SEPs: somatosensory evoked potentials

SN: substantia nigra

SND: striatonigral degeneration

SNr: substantia nigra pars reticulata

SOT: secondary orthostatic tremor

SPECT: single-photon emission computed tomography

STN: subthalamic nucleus

STT: spino-thalamic tract

TCA: Tremor Coherence Analyzer

Thal: thalamus

TMS: transcranial magnetic stimulation

TRS: Tremor Rating Scale

UMN: upper motoneuron

UPDRS: Unified Parkinson Disease Rating Scale

V: vermis

VA: ventral anterior

Vce: ventralis caudalis externa

Vci: ventralis caudalis interna

Vim: ventralis intermedius nucleus

VLN: ventral lateral nucleus

Vm: membrane potential

VOA: ventralis oralis anterior nucleus

VOP: ventralis oralis posterior nucleus

VP: ventral posterior

VSCT: ventral spinocerebellar tract

VST: vestibulospinal tract

VTSS: Vocal Tremor Scoring System

WOTAS: wearable orthosis for tremor assessment and suppression

Xi: concentration of the ion inside the membrane

Xo: concentration of the ion outside the membrane

ZI: zona incerta

Author Biography

Giuliana Grimaldi received her MD degree from the University of Palermo, Italy. She is working on the applications of haptic technology in neuroengineering and rehabilitation, with an emphasis on myoelectric applications. She is a member of the Society for Cerebellar Research (SRC). She is affiliated with the Clinical Neuroscience Department of the University of Palermo, Italy. She is a visiting scientist of the Free University of Brussels, Belgium.

Mario Manto received his MD degree from the Free University of Brussels, Belgium. He has a PhD degree in neurophysiology. He has published more than 90 full papers in refereed journals. He is the founding editor of the international journal *The Cerebellum* and is the associate editor of the *Journal of NeuroEngineering and Rehabilitation* (*JNER*). He is the founder of the Society for Cerebellar Research (SRC). He has served as reviewer for more than 10 international journals. He teaches human physiology at the Faculty of Applied Sciences, Free University of Brussels. He is currently a researcher at the Belgian National Research Foundation (FNRS). He is active in international (European and transcontinental) networks on tremor. He has patented several devices in the field of tremor.

Index

Printed in the United States
by Baker & Taylor Publisher Services